# FISHERMAN'S FAVOU

# FISHERMAN'S FAVOURITES
# Fish, Shellfish and Seaweed Recipes

Edited by Carolynn Bruton and Liz Tarr
Illustrations by Dave Voorvelt

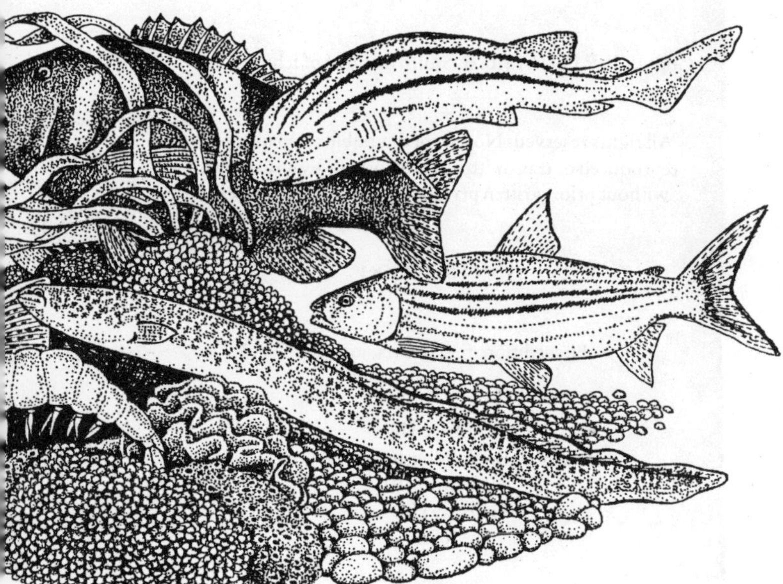

SOUTHERN
BOOK PUBLISHERS

Recipes contributed by well-known ichthyologists, anglers and seafood cooks, compiled by ICHTHOS, the Society of Friends of the J. L. B. Smith Institute of Ichthyology, Private Bag 1015, Grahamstown, 6140



ISBN 1 86812 258 1

First edition, first impression 1992

Published by
Southern Book Publishers (Pty) Ltd
PO Box 3103
Halfway House
1685

Cover design by William Steyn
Illustrations by Dave Voorvelt
Set in 10 on 12 point Bembo
by Book Productions, Pretoria
Printed and bound by Sigma Press (Pty) Ltd, Pretoria

# ACKNOWLEDGEMENTS

Our gratitude goes to those who so generously parted with their favourite recipes and ideas, to Linda Coetzee for her meticulous word-processing, and to Louise Grantham, Mitzi Schutte and Mike Bruton whose considerable editorial skills helped to make this project possible.

# INTRODUCTION

One of the objectives of the J. L. B. Smith Institute of Ichthyology is to contribute to the knowledge and promote the wise use of marine and freshwater fishes. This objective is achieved by means of various research, teaching and curation activities. Our view of natural resource management is that a resource needs to be used in order to be fully appreciated. We therefore actively support the utilisation of fishes by angling, commercial fishing, underwater photography, fish farming, aquarium keeping, etc., as long as this is done in a rational and sustainable way.

One of the great attractions of fishes to humans (and other predators) is that they are delicious to eat. If we are to enjoy their taste to the maximum, we should endeavour to prepare fishes and shellfishes in such a way that we capitalise on their best qualities. A thorough knowledge of fishes and experience of simple ways in which to prepare them for eating are therefore essential qualities of a good seafood cook. In the context of this book we have used the term 'seafood' to cover both marine and freshwater species.

Every year towards the end of September, the J. L. B. Smith Institute of Ichthyology and the Department of Ichthyology and Fisheries Science at Rhodes University host a traditional fish supper for over one hundred guests. In the beginning the students were required to prepare a dish using the fish that they were studying at the time, but since then we have encouraged both staff and students to prepare dishes using any kind of aquatic life, the more unusual the better. Nowadays more than fifty seafood dishes are served at the dinner.

As a result of the outstanding success of the fish supper, we were encouraged to make our recipes available to ICHTHOS members and the general public. This is how the ICHTHOS recipe book was spawned! In addition to our own recipes, we invited various prominent ichthyologists, anglers, fisheries administrators and seafood cooks to contribute their favourite recipes. Many of the recipes are named after their originators and we have endeavoured, as far as possible, to retain the informal, personal style of each contributor. The result is an outstanding collection of tried and trusted recipes based on a wealth of practical experience.

This book includes 289 recipes, all of which use readily available ingredients. We have concentrated on economical and healthy cooking methods, and innovative recipes that wrest the best from simple ingredients. In order to encourage more adventurous cooking, unusual seafoods have been suggested in some recipes, but these may be exchanged for ones that are more readily available from your local supermarket if necessary.

A good menu should consist of dishes that are pleasing to the eye as well as nutritious and filling. The food chosen should provide variation in flavour, texture and colour. The seafood recipes presented here will enable you to serve a truly balanced menu. As the presentation of food is important, we have also provided various ideas for garnishing and dressing the dishes.

Some of the recipes, if made in smaller quantities, can serve as hors d'oeuvres. Seafood also makes excellent starters – light, tasty, attractive and interesting. Nowadays, with good health being uppermost in many people's minds, seafood is regarded as nutritious but not too filling as a main course. Seafood provides proteins to build up the body, carbohydrates to meet energy demands, as well as vitamins, minerals and roughage to regulate essential body processes – in short, everything that is needed for a well-balanced diet. Seafood also tends to be more versatile than traditional red meat and lends itself to many easy-to-prepare dishes that will satisfy the most discerning palate.

People sometimes complain that seafood 'goes off' too easily. This does not happen if it is treated correctly. The most important point is to buy or use only very fresh seafood. To test a whole fish for freshness, look for the following:

- the gills should be firm and red or pink,
- the flesh should be firm and elastic to the touch,
- the eyes should be bright and bulging,
- the scales should be tightly attached to the body,
- the fish should have a fresh odour, and
- it should sink if placed in a bowl of water.

The other important point is that fish should never be refrozen. Before consuming shellfish, check with your local fisheries inspector whether there have been any recent outbreaks of red tide.

Good cooking, and rest assured that part of the proceeds from the sale of this book will be used for fish research and conservation.

Mike Bruton
J.L.B. Smith Institute of Ichthyology

# METRIC CONVERSION TABLES

*These metric quantities are not exact equivalents of the old quantities, but can be more easily measured with the new metric kitchen measures and massmeters.*

## *Volume measurements*

| | | |
|---|---|---|
| ¼ t | = | 1,5 m*l* |
| ½ t | = | 3 m*l* |
| 1 t | = | 5 m*l* |
| 1 T | = | 15 m*l* |

| | | |
|---|---|---|
| 1 pint | = | 600 m*l* |
| 2 pints | = | 1,250 *l* |
| 3 pints | = | 1,900 *l* |
| 4 pints | = | 2,500 *l* |
| 1 gal | = | 4,5 *l* |
| 2 gal | = | 9,0 *l* |

| | | |
|---|---|---|
| ¼ cup | = | 60 m*l* |
| ⅓ cup | = | 80 m*l* |
| ½ cup | = | 125 m*l* |
| ¾ cup | = | 190 m*l* |
| 1 cup | = | 250 m*l* |

*In the UK, the cups to ml conversion differs from the one mentioned above. Here is the UK version.*

| | | |
|---|---|---|
| ¼ cup | = | 80 m*l* |
| ⅓ cup | = | 100 m*l* |
| ½ cup | = | 150 m*l* |
| ¾ cup | = | 230 m*l* |
| 1 cup | = | 300 m*l* |

## *Mass measurements*

| | | |
|---|---|---|
| ½ oz | = | 15 g |
| 1 oz | = | 30 g |
| 2 oz | = | 60 g |
| 3 oz | = | 90 g |
| 4 oz | = | 125 g |
| 5 oz | = | 150 g |
| 6 oz | = | 180 g |
| 7 oz | = | 210 g |
| 8 oz | = | 250 g |
| 1 lb | = | 500 g |
| 2 lb | = | 1 kg |
| 3 lb | = | 1,5 kg |
| 4 lb | = | 2 kg |
| 5 lb | = | 2,5 kg |

## *Oven temperatures*

| | CELSIUS | FAHRENHEIT | GAS |
|---|---|---|---|
| very cool | 110° C | 225° F | ¼ |
| | 120° C | 250° F | ½ |
| cool | 140° C | 275° F | 1 |
| | 150° C | 300° F | 2 |
| moderate | 160° C | 325° F | 3 |
| | 180° C | 350° F | 4 |
| moderately hot | 190° C | 375° F | 5 |
| | 200° C | 400° F | 6 |
| hot | 220° C | 425° F | 7 |
| | 230° C | 450° F | 8 |
| very hot | 240° C | 475° F | 9 |

# CONTENTS

# *Seaweed*

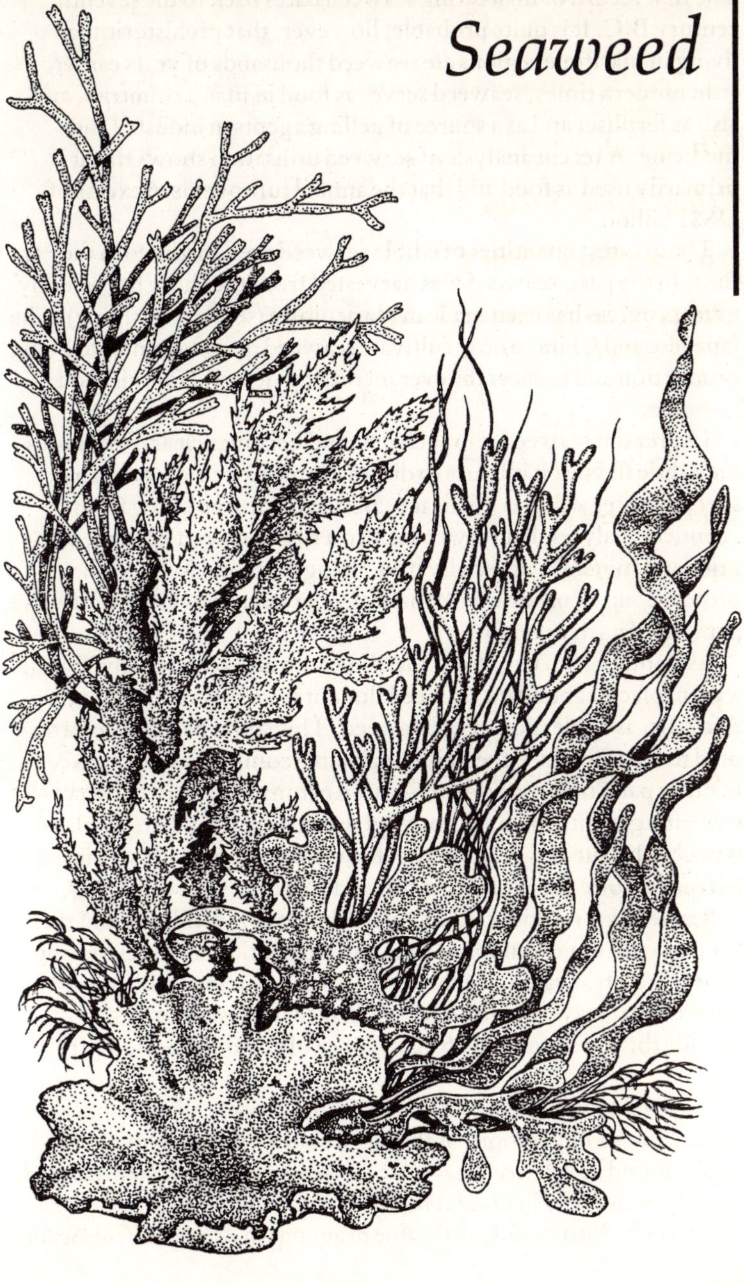

The first record of man eating seaweed dates back to the seventh century B.C. It is quite probable, however, that prehistoric man living in maritime regions ate seaweed thousands of years earlier.

In modern times, seaweed serves as food in many countries, and also as fertiliser and as a source of gelling agents in industry and medicine. A recent analysis of seaweed utilisation shows that it is primarily used as food and that the annual turnover is in excess of US$1 billion.

The greatest quantities of edible seaweed are consumed in the Far East. In the past, seaweed was harvested from the shore but recently great progress has been made in mariculture (seaweed farming). The Japanese and Chinese now cultivate seaweed on rafts or nets to boost production and to meet the ever increasing demand for this food resource.

The reason seaweed is in such demand becomes clear if one tastes the subtle flavour it imparts to dishes. Some extremely palatable species are regarded as a delicacy. Besides taste, we know from chemical analyses that seaweed is also a rich source of protein, vitamins and minerals, especially iron, manganese and iodine. The protein content of seaweed constitutes between four and twenty-five per cent of its dry weight.

As South Africa has a long coastline with an exceptionally diverse and, in places, abundant seaweed flora, it is surprising that South Africans, as a rule, do not eat seaweed. This could perhaps be attributed to two factors: either it is unpalatable compared with seaweed in other parts of the world, or South Africans have very conservative eating habits and have not tried seaweed in their diet. The latter is probably true and we hope that the recipes we provide will inspire the reader to be more adventurous and try some seaweed dishes.

Seaweed is best collected on rocky shores at low tide around the time of a full or new moon. These spring low tides make collecting easier, as a great variety of seaweed is exposed at low tide. It must be stressed that seaweed may not be collected in these reserves; outside these reserves only 5 kg may be collected per person per day.

Seaweed species can be identified from a few excellent publications. Colour illustrations of the more common species of seaweed can be found in *The Seaweeds of the Tsitsikamma Coastal National Park* by S. Seagrief, and in *Our Living Shores* by G.M. Branch and M. Branch. Also useful are the line drawings in *Marine Life on South*

*African Shores* by J. Day, and two articles by S. Seagrief: 'Seaweeds of Maputaland' in *Studies on the Ecology of Maputaland,* edited by M.N. Bruton and K.H. Cooper, and 'Marine Algae' in *A Field Guide to the Eastern Cape Coast,* edited by R.A. Lubke, F.W. Gess and M.N. Bruton.

Mark Aken

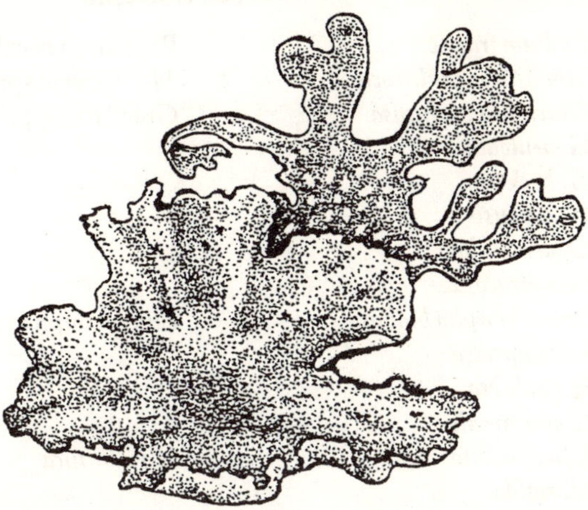

# SEAWEED SPECIES CONVERSION CHART

| *South African species* | *Northern hemisphere equivalent* |
|---|---|
| **Brown algae** | |
| *Laminaria pallida* | *Laminaria longissima* |
| *Ecklonia biruncinata* | *Pleurophycus gardneri* |
| *E. radiata* | |
| *E. maxima* | |
| *Sargassum elegans* | *Ascophyllum mackaii* |
| *S. heterophyllum* | *Sargassum* spp. |
| **Red algae** | |
| *Porphyra capensis* | *Porphyra miniata*<br>*P. tenora, P. suborbiata,*<br>*P. nereocystis* |
| *Gracilaria verrucosa* | Present in northern hemisphere |
| *Epymenia capensis* | |
| *E. obtusa* | *Palmaria palmata* |
| *Rhodymenia natalensis* | |
| **Green algae** | |
| *Codium fragile* | Present in northern hemisphere |
| *Monostroma wittrocki* | *Monostroma* spp. |
| *Gracilaria verrucosa* | *Gracilaria* spp. |
| *G. aculeata* | |
| *G. beckeri* | |
| *G. capensis* | |
| *G. denticulata* | |
| *G. verrucosa* | |
| *Enteromorpha bulbosa* | *Enteromorpha* spp. |
| *E. compressa* | |
| *E. prolifera* | |
| *E. intestinalis* | |
| *Ulva fasciata* | *Ulva lactuca* |
| *U. rigida* | |

# *Ham and* Laminaria *rolls*

(brown algae: *Laminaria pallida, Ecklonia biruncinata, E. radiata, E. maxima*)

10 pieces dried *Laminaria*
10 slices smoked ham *or* bacon
water
¼ cup soy sauce
3 T honey
string *or* toothpicks

Soak the *Laminaria* for 1 hour in enough cold water to cover. Drain and cut into large rectangles. Cut ham to same size. Place a slice of ham on top of each piece of sea vegetable and roll up from one end. Tie with a piece of string or pin with a toothpick. Place in a saucepan. Cover with water. Add soy sauce and honey. Cook uncovered over medium heat until most of the liquid has evaporated and the *Laminaria* is tender (about 30 minutes).

---

# *West coast snack*

(brown algae: *Laminaria pallida, Ecklonia biruncinata, E. radiata, E. maxima*)

10 pieces sea vegetable
½ cup soy sauce
½ cup honey

Rinse sea vegetable in cold water to remove any sand. Spread out in strong sun and wind until completely dry. Store in plastic bags. When ready to use, snip into 30–60 mm pieces or squares and bake at 200° C for 15–20 minutes.

Prepare marinade, mixing well. Place partially toasted strips of sea vegetable in the marinade and toss to coat. Place coated strips on a foil-lined baking sheet and bake at 200° C for 20–30 minutes until crisp. Turn at intervals while baking.

---

# Porphyra *chips*

(red alga: *Porphyra capensis*)

Porphyra *consists of about 20% protein and has a very low calorie content. It is a slippery,* Ulva-*type seaweed found high up on the littorina zone. It is best prepared just before using, as* Porphyra *readily absorbs moisture from the air and is tastiest when crisp. When these delicious chips are ready they are stacked and pinned with toothpicks, otherwise the slightest breeze would float them off the plate.*

Wash fresh *Porphyra* quickly in cold water to remove any sand. Do not soak! Spread out in strong sun and breeze until completely dry. Store in plastic bags. When ready to use, snip into 30–60 mm strips and bake at 200° C for 20–30 minutes until crisp. Eat like potato chips, with tea or cocktails.

---

# Porphyra *soup*

(red alga: *Porphyra capensis*)

1 cup dried crumbled *Porphyra* (see *Porphyra* chips)
1 T salad oil
1 small onion, thinly sliced
½ cup thinly sliced celery
1 t fresh ginger, minced
1 T soy sauce
1 tin beef broth (410 g)
1 cup water

Heat oil in a saucepan over medium heat and sauté onion and celery until onion is translucent (5 – 7 minutes). Stir in the ginger, soy sauce, beef broth, water and sea vegetable. Bring to the boil. Reduce heat, cover and simmer for 5 minutes.

# *Pacific sea vegetable soup*

(red alga: *Porphyra capensis*)

*A well-flavoured fish stock is the secret to this tasty soup. Fish stock can be purchased in cubes. However, it is simple to make and can be stored in the refrigerator* or *freezer. Simply put the head and bones of a fish in enough water to cover, add salt and a little white wine. Boil uncovered until the liquid is reduced by half and strain through cheesecloth.*

4 strips dried *Porphyra* fronds
sesame oil
1 small onion, thinly sliced
1 cup frozen peas
10 mm slice fresh ginger root *or* 2 t powdered ginger
1 T soy sauce
2 cups fish stock, heated

Heat a little oil and sauté the onion until it becomes translucent. Stir in the remaining ingredients and bring to the boil. Reduce heat. Cover and simmer for 5 minutes.

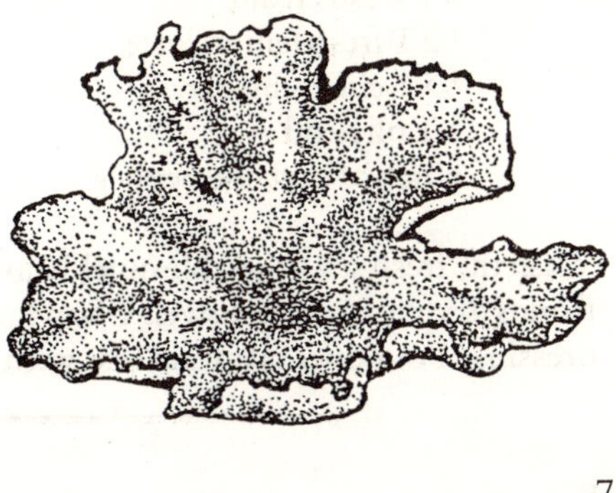

# Ulva *soup*

(green algae: *Ulva fasciata, U. rigida*)

Ulva *makes a nice soup base and background flavour for most clear soups.*

½ cup fresh *Ulva* fronds
2 cups water
½ cup carrots, chopped
½ cup leftover cooked chicken, diced
2 T soy sauce
½ t black pepper

Wash and clean the *Ulva* thoroughly in lukewarm, fresh water removing any small snails that may be feeding on the blades. Boil the carrots until tender. Cut the *Ulva* into 25 mm squares and drop into the boiling liquid. Add the diced chicken. Simmer gently for 5 minutes, season and serve.

---

# *Sea lettuce salad*

(green algae: *Ulva fasciata, U. rigida*)

5 cups fresh *Ulva* fronds
½ cup onion, thinly sliced

## Cream dressing

8 T fresh cream
½ T fresh lemon juice
½ T cider vinegar
1 T olive oil
pinch of cayenne pepper

Wash the fresh sea vegetable quickly in lukewarm water. Pat the fronds dry with a clean towel. Chop finely. Prepare the cream dressing. Pour over salad and toss with the onion. (Serves 2 to 4.)

---

# *Miyok-kuk*

(brown alga: *Endarachne binghamiae*)

*This sea vegetable soup comes from Korea.*

15 g dried *Endarachne*
125 g raw lean beef, thinly sliced
3 cloves garlic, finely chopped
1 t sesame salt
3 T soy sauce
2 t sesame oil
½ t black pepper
6 cups water

Put the *Endarachne* in a colander and blanch by pouring boiling water over it. Drain. Cut into 40 mm pieces. Brown the beef quickly in a little sesame oil with the garlic, sesame salt, soy sauce and black pepper. Add the sea vegetable and the water. Simmer for 30 minutes or until *Endarachne* is tender. Adjust seasoning. Bring to the boil and serve. (Serves 4 to 6.)

---

# *Dulse spread*

(red algae: *Epymenia* spp., *Rhodymenia natalensis*)

1 cup dried dulse
1 cup spring greens, washed and drained

Roast dried dulse in the oven at 250° C until it becomes crisp but not bitter (about 45 minutes). Pulverise the roasted dulse with a mortar and pestle or a mixer. Grind this powder with the chopped spring greens to form a paste. Serve with rice, as a filling for ravioli, or spread on crackers.

---

# *Sea vegetables Taiwan*

(red algae: *Porphyra capensis, Monostroma wittrocki, Enteromorpha* spp.)

1 cup fresh *Porphyra* fronds
1 cup fresh *Monostroma* or *Enteromorpha* fronds
½ cup fresh mushrooms, sliced
1 t honey
2 T soy sauce
pepper to taste
4 T peanut oil

Wash fresh sea vegetables quickly under cold running water. Do not soak! Pat dry with a clean towel. Combine soy sauce, honey and seasoning. Heat oil in a frying pan. Gently sauté the mushrooms until soft. Add sea vegetables and sauté for another 2 minutes. Add seasoning and liquid mixture. Bring to a quick boil. Serve hot. (Serves 2 to 4.)

---

# *Dulse salad*

(red algae: *Epymenia capensis, E. obtusa*)

### Salad
90 g dried dulse
2 apples, sliced
3 large carrots, grated
1 cup alfalfa sprouts
½ ripe avocado, chopped
¼ cup currants

### Dressing
⅔ cup oil
juice of 1 lemon
4 T soy sauce

Chop the dried dulse finely. Mix all salad ingredients together. Mix the dressing and pour it over the salad. Toss well. (Serves 8.)

---

# *Philippine gulamon salad*

(green alga: *Codium fragile;* red alga: *Gracilaria verrucosa*)

3 cups fresh *Codium* or *Gracilaria,* tightly packed
4 large tomatoes
1 t sea salt
½ cup chopped scallions
¼ cup chopped fresh ginger root
2 T soy sauce

Wash and clean the sea vegetable in cold water. Cut off the hard ends of the *Codium* plants, if used. Blanch by placing the sea vegetable in a colander and pouring boiling water over it. The *Gracilaria* will change from a brownish-red colour to brownish-green. Allow to stand for a few minutes. Drain well. Chop or mash the tomatoes and combine with the blanched sea vegetable. Add the remaining ingredients. Mix gently. Serve cold. (Serves 8 to 10.)

---

# *Hawaiian* Sargassum

(brown algae: *Sargassum elegans, S. heterophyllum*)

1 cup fresh *Sargassum* (tender young leaves only)
1 T sea salt

Wash the *Sargassum* well in fresh water, removing all sand, bits of coral or clinging algae. Remove the leaves from the stipes and soak them overnight in cold fresh water. Drain. Use as a stuffing for baked fish.

The *Sargassum* blades may also be used without soaking. Chop finely or grind, salt and refrigerate for future use in soups, etc. Salted, it will keep for long periods.

---

# *Ogo kim chee*

(red algae: *Gracilaria* spp.)

*This pickled ogo dish is prepared in Korean coastal communities all over the world.*

1 kg fresh ogo (*Gracilaria* spp.), chopped into 30–60 mm pieces
½ cup salt
2 cloves garlic, chopped
1 to 2 onions, chopped
1 t chopped red chillies *or* ½ t cayenne pepper
½ t paprika

Wash and clean the ogo. Sprinkle with the salt and allow to stand overnight to wilt. Next day, drain off any liquid that has formed. Add garlic, onions, chillies (or cayenne pepper) and paprika. Toss to coat the ogo. Pack tightly into jars. Seal and refrigerate. Let stand for a few days before using.

# Shellfish

# PRAWNS

## *Schoonbee's king prawns*

12 king prawns
2 cups dry white wine
100 g butter
2 chopped onions
1½ lemons
2–3 cloves crushed garlic
150 g finely chopped mushrooms
seasoning

Cut open tails of prawns mid-dorsally. Remove thin black intestine and stomach in head region. Wash prawns under a tap and dry. Do not remove head portions. Melt butter in pan. Add wine, onion and garlic. Rub some onion and garlic mixed with salt into cut-open tails. Place prawns in butter, onion and garlic sauce. Cook for 5 minutes. Turn prawns over and cook for another 5 minutes. Squeeze juice of half a lemon over prawns. Cook for 2 minutes more. Remove prawns from pan and keep warm. Add seasoning to sauce and then add mushrooms. Cook the sauce for 3–5 minutes. To serve, put prawns on a bed of rice and pour prepared sauce over. Decorate with a thinly sliced lemon and chopped parsley. Serve hot.

# *Harry Voigt's prawn batter and sauce*

12 prawns

### Batter

2 T flour
½ t mustard
salt and pepper
2 eggs
3 cloves garlic, finely chopped
milk

### Sauce

2 t sugar
1 clove garlic, finely chopped
1 lemon
1 T ginger
1 small onion, chopped
salt and water

Combine ingredients for batter and mix until creamy. Boil sauce until syrupy. Coat prawns in batter, fry and serve with sauce.

---

# *Classic curried prawns*

500 g cooked, shelled prawns
¼ cup butter
⅓ cup chopped onion
¼ cup chopped green pepper
3 T flour
1 clove garlic, minced
1 t curry powder
2 cups cream
2 T lemon juice
½ t salt
⅛ t each pepper, ground ginger and chilli powder

Melt butter over medium heat, add onion and green pepper and cook for 5 minutes. Add flour, garlic, curry powder and stir until bubbly. Remove from stove, beat in cream, lemon juice, salt, ginger, pepper and chilli powder. Return to stove and cook over medium heat until thick. Add prawns and heat thoroughly. Serve at once on a bed of rice. (Serves 6.)

---

# *Piraeus feta-baked prawns*

1 kg cooked, shelled prawns
1 onion, finely chopped
3 T oil
1 cup water
juice of 2 lemons
2 × 250 g tins tomato paste
1 T cornflour dissolved in 1 T water
½ t origanum
100 g feta cheese cut into cubes (*or* substitute cottage cheese)

Preheat oven to 350° C. Sauté onion in oil until soft. Add water, lemon juice and tomato paste and bring to the boil. Stir in cornflour mixture and simmer until thick. Add prawns and origanum, stirring well. Mix in feta cubes and place in flameproof dish. Bake uncovered for about 15 minutes or until cheese melts. (Serves 4.)

---

# *Grapefruit and prawn cocktail*

1 tin grapefruit segments (500 g)
100 g cooked prawns
small bunch of white grapes (tinned grapes may be substituted)
2–3 T mayonnaise
cayenne pepper
a few prawns for garnishing

Drain the syrup from the grapefruit segments and cut into smaller pieces. Mix with prawns, grapes (peeled and seeded) and mayonnaise. Pile into cocktail glasses and decorate with a few fresh prawns. Sprinkle with a dash of cayenne pepper. (Grapes can be popped into boiling water for a few seconds to make peeling easier.)

---

# *Portuguese prawn croquettes*

575 g fresh prawns, unpeeled
salt and white pepper
25 g butter
60 m*l* flour
2 egg yolks
1 egg
100 g dry breadcrumbs
450 m*l* olive oil

Wash the prawns thoroughly. Put them in a saucepan with 1 t salt, cover with water and boil for 5 minutes. Remove the prawns with a slotted spoon and peel when cool enough to handle. Return the trimmings to the boiling water and boil for a further 10 minutes. Strain and reserve the stock. Once the prawns have cooled, chop them finely. Melt the butter in a saucepan, sprinkle with the flour and stir until smooth. Gradually beat in 150 m*l* of the reserved stock and bring to the boil over a low heat. Remove from heat. Stir in the chopped prawns and egg yolks and season. Heat through quickly, remove from heat and set aside to cool. When the mixture has cooled, shape it into small balls. Beat the egg, dip the balls in the egg, then coat with breadcrumbs. Heat the oil and fry the croquettes until brown and crisp all over. Drain on absorbent kitchen paper and serve hot. (Serves 4.)

---

# *Ofer's Pacific fried prawns*

12 prawns, fresh *or* frozen
½ t salt
1 T soy sauce
1⅓ cups milk
1 cup shredded coconut
¼ cup lemon juice
1½ t curry powder
¼ t ground ginger
2¼ cups flour
2 t baking powder
cooking oil

Remove shell from tail section of prawns. Mix lemon juice, salt, curry, ginger and soy sauce. Pour over prawns and marinate for at least one hour. In a shallow bowl, roll prawns in ¼ cup of flour. Mix the rest of the flour, milk, baking powder and remaining marinade into a smooth batter. Preheat oil for deep frying. Dip prawns in batter and drop into hot oil. Fry until golden brown. (Serves 4.)

---

# *Ofer's Cantonese prawns*

5 prawns
1 *l* water
½ cup chicken stock
½ t salt
¼ t sugar
1 T oil
1 T white wine
1 T soy sauce
1 T thinly shredded ginger root *or* 1 t ground ginger
2 spring onions

Remove claws and legs. Slice in half and remove dark vein if visible. Boil water and drop prawns in for 1 minute. Drain water and set prawns aside. Chop spring onions in 4 cm pieces. Preheat wok or large pan to a high temperature and add oil. Add onions, ginger, chicken stock, wine, soy sauce, salt and sugar while stirring. Add prawns and cook for 1 minute, stirring constantly. (Serves 2.)

---

## *Ofer's Hawaiian prawns*

12 prawns
1 clove garlic, crushed
1 T onion, finely chopped
2 T parsley
pinch salt and a dash of pepper
¼ cup white wine
4 T oil

Remove claws and legs. Heat oil in a large frying pan. Add garlic and onion and fry for two minutes. Add prawns and sauté for 4 minutes, flipping over often. Add wine, salt and pepper and cook for another 2 minutes while stirring. Sprinkle with parsley and serve with warm butter. (Serves 4.)

---

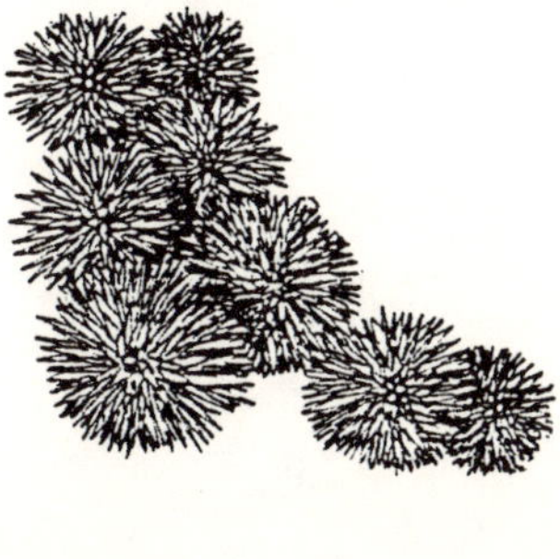

# SHRIMP

## *Shrimp salad*

1 cup diced apple
1 cup tinned pineapple (reserve juice)
180 g cooked shrimp, coarsely chopped
½ cup sour cream
cottage cheese and unpeeled orange segments to garnish

Mix 2 T pineapple juice with sour cream, toss with the rest of the ingredients and spoon into a serving bowl. Garnish with cottage cheese and orange segments.

---

## *Shrimp dip*

Mince 180 g cooked shrimp. Add 600 m*l* sour cream mixed with one packet dry onion soup. Serve with chips or crudités.

---

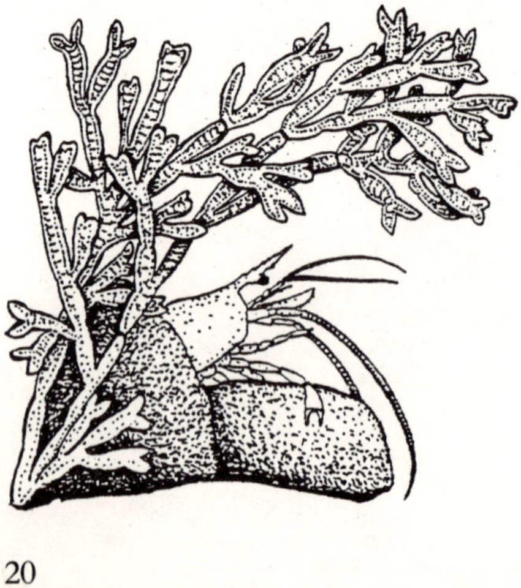

# *Shrimp in hot sauce*

900 g large shrimps, shelled
50 m*l* butter, melted and strained
1 medium onion, chopped
2 cloves garlic, crushed
3 cm strip fresh ginger, peeled and chopped
5 m*l* turmeric
5 m*l* hot chilli powder
12,5 m*l* ground coriander
5 m*l* ground cumin
5 m*l* salt
2 m*l* black pepper
140 m*l* yoghurt

Melt the butter in a saucepan. Add the onion, garlic and ginger and fry, stirring occasionally, until the onion is golden brown. Add the shrimps and fry for 5 minutes. Stir in the spices, seasoning and yoghurt. Bring to the boil, reduce the heat to low and simmer for 30 minutes, or until sauce has thickened. Stir frequently. (Serves 6.)

---

# *Shrimp soufflettes*

1 tin small shrimps (200 g)
15 m*l* gelatine, dissolved in a little hot water
250 m*l* chicken stock (½ cube chicken stock dissolved in 250 m*l* warm water)
250 g smooth cottage cheese
12 m*l* anchovy paste
1 m*l* dried dill
5 m*l* onion, finely grated
125 m*l* cream, whipped
pinch of sugar
2 large egg whites, stiffly beaten

Mix cheese, anchovy paste, dill and onion with a wooden spoon. Stir gelatine into cheese mixture and add cooled stock. Beat well. Refrigerate to the point of setting. Put shrimps into sieve, rinse well, drain and pat dry. Reserve 6 shrimps for garnishing and add the rest to the cheese mixture. Fold in cream, egg whites and sugar. Spoon into six lightly oiled or rinsed ramekins, or a single bigger container. Chill until firm. Unmould on a perfect lettuce leaf. Top with reserved shrimps. (Serves 6.)

# *Shrimp starter in shells*

250 g cooked prawns *or* shrimps
300 m*l* thick white sauce
4 hard-boiled eggs, chopped
1 cup grated cheese
2 t chopped parsley
salt, pepper and cayenne pepper

Add chopped eggs, cheese, shrimps and parsley to white sauce. Heat gently and season to taste. Pour into individual shell dishes. Sprinkle grated cheese on top and brown under grill. Serve hot.

## *Lien's crayfish*

3 large crayfish, cooked
125 m*l* white wine
125 m*l* water
1 small onion, sliced
1 stalk celery
1 sprig parsley
salt and pepper
mushrooms, sliced
1 tin pineapple pieces (500 g)

Shell and cube crayfish. Put crayfish, wine, water, onion and seasoning in a greased ovenware dish. Cover with greaseproof paper and poach at 175° C for 20 minutes. Allow to cool in liquid. Reserve liquid for sauce.

### Sauce

250 m*l* mayonnaise
60 m*l* cream
60 m*l* tomato sauce
1 m*l* grated onion

Mix sauce ingredients together and fold in cubed crayfish, reserved liquid, mushrooms and pineapple. Refrigerate. Serve on a bed of lettuce.

# *John's genuine crayfish*

*You have never really tasted crayfish until you have tasted it like this!*

Catch several good-sized crayfish. Make a fire with driftwood and have a beer while the coals mature. When the coals are ready, place the crayfish, back down, directly on to the coals. Sprinkle with seawater or beer from time to time to prevent them from burning. Turn once, briefly, when flesh is almost cooked. Remove from coals and have a beer while allowing crayfish to cool down slightly. Eat with fingers, with beer as a side dish.

---

# *Crayfish cocktail*

500 m*l* cold, cooked crayfish
250 m*l* mayonnaise
125 m*l* tomato sauce
10 m*l* finely grated onion
5 m*l* chutney
10 m*l* Worcestershire sauce
15 m*l* lemon juice
salt and pepper

Mix all the ingredients (besides the crayfish) well. Dice crayfish, put in a dish and coat with mixture. Refrigerate. Line stemmed glasses with lettuce and spoon crayfish mixture into each glass.

---

# *Savoury crayfish*

crayfish, cooked
butter
salt and cayenne pepper
125 m*l* white wine

## Sauce

300 m*l* cream
1 egg yolk

Cut crayfish into small pieces and heat slowly in butter – do not allow to brown. Season with salt and cayenne pepper. Add wine and simmer slowly until liquid has been reduced by half. While crayfish is simmering, make a sauce by slowly heating the cream with the egg yolk, stirring constantly. When sauce begins to thicken, remove from heat and add crayfish mixture. Season and serve on fingers of toast.

---

# *Dr Courtenay-Latimer's crayfish soufflé*

Prepare a white sauce by blending 3 tablespoons of butter with 3 tablespoons of flour in a saucepan over low heat. Gradually stir in 1 cup of milk, season with salt and a little cayenne pepper, add a squeeze of lemon juice and half a teaspoon of chopped parsley. Remove from the stove and add the beaten yolks of 3 eggs and a cup of finely shredded or minced crayfish meat. Mix well. Lastly fold in the stiffly beaten whites of 3 eggs. Pour into a soufflé dish and bake in a moderate oven (180° C) for 20 minutes or until set.

---

# CRAB

## *Len Compagno's boiled crab*

2 large crabs
1 bay leaf
1 pinch dill
2 T lemon juice
5 black peppercorns
1 t salt
sufficient rapidly boiling water to cover crabs

Put ingredients in a large pot, bring to the boil and simmer for 25 minutes. Discard the gills and gravel sack. The crab may then be served with either garlic butter or, if allowed to cool, with the meat removed and mixed with fresh mayonnaise and finely chopped chives. A bland green salad with a French dressing, wedges of rye bread and a dry pinotage or cabernet rosé make a perfect accompaniment to this dish. (Serves 4.)

## *French crab soufflé*

180 g crabmeat, flaked
30 g butter
30 g flour
250 m*l* milk
3 eggs, separated
pinch of paprika
salt and pepper

Melt butter in a large pan, slowly adding flour to make a roux. Mix in milk gradually, stirring all the time, to make a thick sauce. Cool slightly, then stir in the beaten egg yolks, salt, pepper and paprika. Add crab meat and fold in stiffly beaten egg whites. Pour into buttered soufflé dish and bake in centre of oven at 180° C for 25 minutes until risen and golden brown. Serve immediately. (Serves 4.)

## *Crab savoury*

1 tin crab meat (300 g)
2 t gelatine
¼ cup hot water
¾ cup milk
2 egg yolks
1 t salt
1 t mustard
2 T melted butter
2 T vinegar
cayenne pepper

Chop crab finely. Dissolve gelatine in hot water. Mix egg yolk, vinegar, butter, milk and seasonings, put in double boiler and cook until mixture begins to thicken. Remove from heat, allow to cool, add gelatine and crab. Place in a shallow mould and put in refrigerator to set. Slice and serve on biscuits or thinly sliced buttered bread.

# *Dressed crab*

1 crab, cooked
2 T vinegar
2 T oil
salt
white pepper
cayenne pepper to taste

Remove the meat from the shell and mix thoroughly with the other ingredients. Rinse the carapace (back shell) and pile the mixture into it. Garnish with slices of lemon and parsley.

---

# *Crab cakes*

500 g crab meat
2 slices bread, crusts removed
milk
1 egg, beaten
1/4 t salt
1 t seasoning
1 t baking powder
1 T chopped parsley
1 T Worcestershire sauce
1 T mayonnaise *or* salad dressing

Break bread into small pieces and moisten with milk; combine with remaining ingredients. Shape into patties. Fry quickly in hot oil until brown.

---

# MUSSELS

*Mussels freeze well and can be a useful addition to many seafood dishes. Prepare as in 'Hoke's SA Steamed Mussels', cool, package and freeze for up to 2 months.*

## *Hoke's South African steamed mussels*

Pour 1 cm water into a saucepan or pot. Fill the pan with mussels, washed and cleaned of external sand and algae, replace the lid (or cover with aluminium foil), and steam until the top mussels are opened and their meat firm. When cooked, open the shell, pluck out the meat and remove the gritty 'beard' with which the mussels attach themselves to rocks. Dip the mussels in melted garlic butter with mixed herbs added for taste.

---

## *Colin Buxton's mussel patties*

50 fresh white mussels (*Donax serrata*)
2 eggs
1 cup flour
100 g butter *or* margarine *or* 2 T oil
1 fresh lemon
salt and freshly ground black pepper
crushed garlic *or* garlic flakes

Open the mussels with a knife and wash them under running water. This is best done by gently squeezing the viscera. Beat the eggs, adding salt, ground pepper and garlic to taste. Melt the butter in a large frying pan. Take 2–4 mussels, dip in the egg, then roll into a patty and coat with flour. Fry over a medium heat until golden brown, garnish with lemon juice and serve immediately.

---

# *Hoke's mussel chowder*

Should you by some miracle have steamed mussels left over from a South African 'mussel bake', they can be put to excellent use in a New England style chowder. All you need is a covered pot and a cutting board. Cook 3–4 strips of bacon until crisp. Remove the bacon rind but retain sufficient fat in which to sauté about 3 medium-sized chopped onions. Once the onion is cooked, add about 1 cm of water to the pot as well as 3–4 medium to large carrots (grated, but not too finely), and 4–5 cubed, medium-sized potatoes. Steam the carrots and potatoes until done, then add the bacon (crumbled), at least one cup (hopefully 2) of steamed mussels, 3 litres of milk and half a cup of cream. Season to taste with salt, black pepper, mixed herbs and thyme. Bring the chowder to a low boil, simmer for 7–10 minutes and serve hot.

This recipe will serve 4–6 hungry fishermen, and should keep for 4–5 days when refrigerated. About half a cup of liquid from the steamed mussels will add more flavour, but this is not essential. Canned mussels, clams or oysters could be substituted.

---

# *Baked mussel omelette*

1 cup cooked mussels
2 T butter
2 T flour
½ t salt
¼ t pepper
1 cup milk
5 eggs
¼ cup water
1 t salt
2 T butter
¼ cup grated cheddar cheese

Melt butter in a saucepan; stir in flour, salt and pepper. Slowly pour in milk, stirring constantly, until sauce thickens. Stir in mussels and simmer until heated through. Combine eggs, water and salt and beat well. Melt butter in a frying pan, then pour in the egg mixture. Cook until bottom is lightly browned and centre has just set. Then roll up and place in a baking dish. Sprinkle with cheese, then pour the sauce over the cheese. Bake at 180° C for 8–10 minutes. (Serves 4.)

---

# *Creamed mussels on toast*

2 cups cooked mussels
50 m*l* butter
1/4 cup flour
2 t lemon juice
1 t Worcestershire sauce
1/4 cup spring onions, chopped
1/2 cup mushrooms, sliced
2 cups milk
1/2 t salt
1/4 t pepper
4 slices bread

Melt butter, blend in flour, lemon juice and Worcestershire sauce. Add spring onion and mushrooms and sauté until tender. Slowly add milk, stirring until slightly thickened. Add mussels and stir until hot, then season with salt and pepper. Toast bread and cut each slice diagonally twice to make toast triangles. Serve mussels rimmed with toast and garnished with spring onions. (Serves 4.)

---

# *Herb-steamed mussels*

1 kg mussels in shells
1 t thyme
1 T chopped parsley
1 lemon, thinly sliced
1 raw carrot, peeled and shredded

Clean mussels, place in a large pot and steam until the shells open (5–7 minutes). Drain broth into a saucepan, add thyme and parsley, cover and simmer for 10 minutes. Place mussels and lemon slices in a bowl. Pour in the broth mixture. Garnish with carrot.

---

## *Baked mussels*

3 cups cooked mussels
1 t grated onion
1 clove garlic, minced
1 T chopped parsley
½ t paprika
½ t origanum
⅓ cup Parmesan cheese
½ t salt
¼ t pepper
½ cup dry breadcrumbs
50 m*l* melted butter

Spread mussels in a greased baking dish. Mix onion, garlic, parsley, paprika, origanum, cheese, salt and pepper. Sprinkle over mussels. Top with crumbs and drizzle with melted butter. Bake uncovered at 200° C for 20 minutes. (Serves 4–6.)

---

## *Rob's grilled mussels*

10–12 fresh mussels per person *or* substitute tinned mussels in shells
500 m*l* water
salt
1 clove garlic, finely chopped
garlic salt
25 g butter
20–40 m*l* seasoned crumbs (home-made *or* bought)
15–20 m*l* lemon juice

Boil fresh mussels in salt water for 5–10 minutes. Discard any which do not open. Remove one half of the shell, leaving the mussel in the other half. (Tinned mussels should not be boiled, but half the shell should be discarded). Mix the chopped garlic, garlic salt and butter and add some to each half shell containing a mussel. Sprinkle with seasoned crumbs and lemon juice and grill for 2–3 minutes. Serve hot.

---

# PERLEMOEN

*There are many ways of preparing and cooking perlemoen: some prefer to cook it in a stock of half wine and half water in a pressure-cooker for about 20 minutes, others to simmer the shellfish in the same quantity of stock for about 1½ hours, to obtain the same result. Perlemoen can also be fried or pressure-cooked or simmered until tender, then minced and made into rissoles.*

## *Do's and don'ts*

1. Don't freeze perlemoen before cooking as freezing will always make it tough.
2. Don't over clean – simply remove gut and brush very lightly to remove outside dirt. Then beat.
3. Don't add salt or water to pot when cooking perlemoen.
4. Kill slowly by soaking in fresh water.

## *Chris Heineken's perlemoen*

*This recipe can also be made with whatever shellfish is available.*

2–3 perlemoen
1–2 cloves garlic, chopped
1 t salt
1 T milk
herbs to taste

Clean perlemoen and beat well. Cut into 2 cm blocks. Place in pressure cooker. Add one level tablespoon margarine or butter for each perlemoen. Heat until margarine has melted. Stir well. Return lid of pressure cooker and seal. Once pressure has built up, time for 15–18 minutes, depending on mass of perlemoen. Switch off heat and allow to cool slowly. Once pressure has been released, the lid can be removed. Mix a little flour and water in a cup and add to the natural juices in the pot. This sauce may be improved by adding a few chopped mushrooms and a tablespoon of chutney.

# *Charles Atkins's scallops of perlemoen in port*

200 g scallops of tenderised perlemoen
25 m*l* port
125 m*l* cream
seasoning

Season scallops of perlemoen and dip in flour. Fry in hot clarified butter for one minute on each side. Add port and cook to reduce. Add cream and reduce further to coating consistency. Adjust seasoning. Place scallops on pre-heated plate on savoury rice. Coat with sauce and serve.

---

# *Allan Heydorn's perlemoen à la seaweed*

4 perlemoen
2 freshly cut pieces of kelp

Remove perlemoen from shell, clean and rinse in vinegar water to remove slime. Clean foot, slice and dice, the smaller the better. Pounding is not necessary.

Cut two large, living kelp stems about 40 cm from frond end so that you have two hollow kelp tubes with the wider portion supporting the fronds. Cut off the fronds. Pack the diced perlemoen into the tubes and seal the open end with plugs of rolled fronds. Do not add salt or spices to the perlemoen, only a little white wine if you wish.

Make a fire and place the kelp stems containing the perlemoen on the coals when most of the flames have died down. Cook in this manner for about 15–20 minutes until the kelp is thoroughly charred. (Drink wine or beer while waiting). Turn the stems occasionally but don't pierce them with a fork.

Remove the stems from the fire and lay them on a flat surface. Cut the stems longitudinally with a sharp knife. The stems will open up to expose the cooked perlemoen. Scoop the perlemoen on to a bed of cooked rice and now add salt according to taste. The perlemoen will have a delicate flavour, having steamed in the juice of the kelp.

Eat and enjoy. A good white wine with the meal is strongly recommended.

---

# *Perlemoen and wine casserole*

*One of the nicest ways of serving perlemoen is in this wine casserole.*

- 2 medium-sized perlemoen, sliced about 7 mm thick, then lightly beaten and cut in half
- 2 T oil
- 1 large onion, thinly sliced
- 1 clove garlic, crushed
- 250 m*l* dry white wine
- 4 rashers streaky bacon, cut into 30 mm pieces
- 1 T chopped parsley
- salt and freshly ground black pepper
- a pinch allspice
- 1 bay leaf
- ½ chicken stock cube
- chopped parsley and chives to garnish

Having prepared the perlemoen, heat the oil in a frying pan and fry the onion and garlic until transparent. Add the perlemoen and fry lightly on each side. Transfer to an ovenproof casserole. Add the wine, bacon, parsley, seasoning and stock cube and bake at 175° C for 1½ hours or until the perlemoen is tender. Should the liquid evaporate during cooking, a little more wine or water must be added to prevent the casserole from burning. Sprinkle with freshly chopped parsley and chives and serve with rice.

---

# Octopus and Squid

## How to prepare fresh or frozen octopus

*Thaw octopus if frozen. If it is whole, slit open head cavity and discard interior. Cut away and discard beak (between legs). Wash octopus thoroughly. Separate legs and head and put in a large pot filled with enough boiling salted water to cover. Simmer gently until skin can be peeled or stripped from flesh (usually about 30 minutes). Let octopus cool in cooking water.*

# *Liz's octopus Port St Johns*

1 medium to large octopus
3 ripe, firm tomatoes, chopped
1 medium-sized onion, chopped
1 green pepper, chopped
1 small, firm avocado, diced
2 cloves garlic, finely chopped
1 T chopped parsley
1 T fresh origanum *or* marjoram, chopped
1 small tin black olives in olive oil and vinegar (300 g)
2 t olive oil
salt and black pepper

Clean and rinse the octopus. Pressure cook in salted water for at least 20 minutes, or until tender. Cool and chop into bite-sized pieces. You may or may not discard the (now purple) skin and suckers – it's a matter of choice. Drain the olives, reserving 2 tablespoons of the liquid for dressing. Mix all ingredients carefully, so as not to break up the avocado, adding salt and liberal sprinklings of black pepper. Mix 2 teaspoons olive oil and 2 tablespoons tinned liquid and gently toss the mixture. Serve chilled with lemon wedges, crusty bread and chilled wine.

---

# *Octopus stew*

2 cups precooked, diced octopus
2 T salad oil
½ cup spring onions, thinly sliced
2 medium-sized tomatoes, peeled and diced
1 T finely chopped parsley
1 clove garlic, minced *or* pressed
¼ t thyme leaves
⅛ t dried rosemary
¾ cup dry white wine
salt

Preheat oven to 160° C. In a large frying pan, heat oil over medium to high heat. Add onion, tomatoes, parsley, garlic, thyme and rosemary. Sauté, stirring occasionally, until vegetables are soft. Add octopus and wine and stir to blend. Pour mixture into a shallow baking dish; bake, uncovered, for 2 hours, stirring occasionally. Add salt to taste and serve hot.

# *Octopus soup*

1 octopus
1 medium-sized onion, sliced
2–3 tomatoes, peeled and sliced
oil
2 T chopped parsley
1 ½ *l* water
salt and pepper
bread slices

Fry onion and tomato in 2 tablespoons of oil. Clean and cut the octopus into small pieces and add to the onion and tomatoes. Fry together for 10–15 minutes before adding parsley and water. Simmer for 2–3 hours. Season to taste. Fry several slices of bread per person, drop them into the soup, cook for another 15 minutes and serve.

---

# *Rob's deep-fried octopus cocktail*

1 kg octopus tentacles
50 m*l* water
50 m*l* cake flour
black pepper to taste
garlic salt to taste
oil for deep frying

## Sauce

½ cup tomato sauce
½ cup mayonnaise
25 m*l* gin
5 m*l* chopped capers
10 m*l* chopped parsley

Boil octopus tentacles in water for 2–3 hours. Add more water if necessary. Allow octopus to cool and then clean (keep the suckers as they are very tasty). Cut the cleaned tentacles into bite-sized pieces. Mix the garlic salt, black pepper and flour in a plastic bag. Add the octopus (and suckers) and coat with the seasoned flour by shaking the bag. Deep-fry the octopus until golden brown. Mix the sauce ingredients and pour over the octopus. This can be served hot or cold. The sauce can also be kept separate and used as a dip for the octopus pieces.

---

# *Tom's spicy squid rings*

This recipe requires 4 fresh squid. Squid is fresh only if still transparent. Clean and skin the squid and cut into rings.

Beat 2 eggs with flour to make a thick batter, add ½ t salt and grind in pepper to taste. Drop the squid rings into the batter. Take a large, heavy pan, toss in 2 tablespoons butter, 2 or 3 squirts of Tabasco sauce and ¼ t chilli paste. Add 1 tablespoon oil and, when sizzling, fry the squid in single layers for not more than ½ minute on each side.

Serve with olives and chilled dry white wine for a finger supper.

---

# *Italian squid*

1½ cups raw squid, sliced
1 large onion, chopped
1 clove garlic, minced
½ green pepper, chopped
3 T oil
1 tin whole peeled tomatoes (300 g)
½ t each basil, chervil, salt
¼ t each rosemary and pepper
2 T chopped parsley
250 g spaghetti

In a large pan fry onion, garlic and pepper in oil until onion is glazed but not brown. Add liquid from tinned tomatoes. Chop tomatoes and add to pan with basil, chervil, salt, pepper, rosemary and parsley. Simmer gently, uncovered, for 20 minutes, until slightly thickened. Add squid and simmer for 8–10 minutes. Serve on spaghetti which has been cooked in boiling salted water until *al dente*. (Serves 4–5.)

---

# *Roman squid and clam insalata*

1½ cups raw squid, sliced into 25 mm rings
1 tin baby clams (250 g)
peel of 1 lemon, cut into very fine strips
¼ cup lemon juice
⅓ cup oil
2 T chopped parsley
salt and pepper

Boil enough water to cover squid. Drop the squid rings into the water; remove them as soon as it starts to boil again, drain and set aside. Drain clam liquid into saucepan and set clams aside. Add lemon peel to clam liquid and bring to the boil. Remove from heat and add squid and clams. Chill in refrigerator. Stir in lemon juice, oil, parsley, salt and pepper, mixing well. Serve chilled. (Serves 4.)

---

# *Oriental stir-fried squid*

1½ cups raw squid, cut into 20 mm strips
3 T oil
2 cloves garlic, minced finely
2 t ginger root, minced finely
125 g mushrooms, sliced
1 medium onion, sliced
1 cup celery, sliced
1 green pepper, sliced

## Sauce

1 ½ T cornflour, blended with 2 T soy sauce
1 t each of sugar and vinegar
½ cup beef stock

Prepare sauce ingredients and set aside. In a wok or large pan, heat 2 T oil and stir-fry garlic, ginger and onion for 1 minute. Add mushrooms and stir-fry for 2 minutes. Remove mixture from pan and set aside. Heat remaining oil in pan. Add squid and stir-fry for 1 minute or until pieces curl. Add celery and pepper, stir-frying for 2 minutes. Add mushroom mixture and sauce mixture and stir until sauce thickens. Serve immediately. (Serves 4.)

---

# *Tom's gensan squid*

1 ½ kg squid
1 t fennel seed
½ t black peppercorns
salt
juice of 2 lemons
olive oil
2 medium heads of fennel
1 large onion
3 stalks celery
5 cloves garlic
60 g chopped parsley
500 g spinach
1 bay leaf
250 g potatoes, thinly sliced
fish stock *or* dry white wine

Clean the squid. Slice into rings and cut tentacles into pieces. Pound peppercorns, fennel seed and salt. Add the lemon juice and an equal amount of olive oil. Pour over the squid, mix and leave to stand while preparing spinach.

Chop fennel, onion, celery and garlic. Put in a large saucepan with half the chopped parsley and the spinach. Pour in some olive oil, then add squid and other vegetables. Season with salt, pepper and a bay leaf. Cook over a low heat for 30–40 minutes. After 15–25 minutes (depending on total cooking time), add potatoes and a little fish stock or wine if squid looks too dry.

---

# *Calamari stew*

1 kg sliced calamari
5 m*l* salt
12,5 m*l* butter *or* margarine
500–750 m*l* red wine
4 tomatoes, chopped
ground pepper to taste
20 m*l* lemon juice
2 cloves crushed garlic
3 m*l* marjoram
2 onions, chopped
37,5 m*l* tomato paste *or* purée

Fry garlic and onion in margarine until well done. Add tomatoes and cook for another 5 minutes. Season calamari with salt and pepper and sprinkle with lemon juice. Add calamari and tomato purée to tomato mixture. Stir until it bubbles, then add wine and marjoram (the calamari should just be covered). Simmer slowly for about 1½ hours. Serve with salads.

---

# *Squid with spaghetti*

1 kg squid, fresh *or* frozen
1 clove garlic, crushed
2 T cooking *or* olive oil
½ t salt
1 tin tomatoes (300 g)
2 T chopped parsley
½ t origanum
1 packet of spaghetti (500 g)

Clean squid. Cut into 2,5 cm pieces. Cook in a heavy saucepan for 5 minutes, stir occasionally. Drain. Cook squid and garlic in hot oil for 5 minutes. Press tomatoes through a colander or liquidise. Add tomatoes, parsley, origanum and salt; stir. Cover and simmer for about 10 minutes or until squid is tender. Serve on spaghetti. (Serves 4–6.)

---

# *Sweet and sour squid*

1 kg whole squid, fresh *or* frozen
1 cup chopped celery
1 cup chopped onion
¼ cup cooking *or* olive oil
2 T drained capers
¼ cup tomato paste
¼ cup cider vinegar
1 T sugar
½ t salt
⅛ t pepper

Clean squid and cut into 2,5 cm strips. Cut tentacles into 2,5 cm pieces. Cook celery and onion in hot oil until tender. Add squid. Cover and simmer for 5 minutes. Rinse and drain capers. Add capers, tomato paste, vinegar, sugar, salt and pepper. Cover and simmer for about 20 minutes or until squid is tender, stirring occasionally. (Serves 3–4.)

---

# *Fried squid*

1 kg whole squid, fresh *or* frozen
2 T lemon juice
1 t salt
⅛ t white pepper
1 egg, beaten
2 T milk
1 cup flour
fat for frying

Clean squid and cut into pieces. Sprinkle with lemon juice, salt and pepper. Combine milk and egg. Dip squid in milk and roll in flour. Put a single layer of squid in a frying pan with hot oil. Fry at moderate heat for 3–5 minutes. Turn carefully and fry for another 3–5 minutes or until squid is lightly browned. Drain on absorbent paper. Serve with lemon wedges.

# Shark and Skate

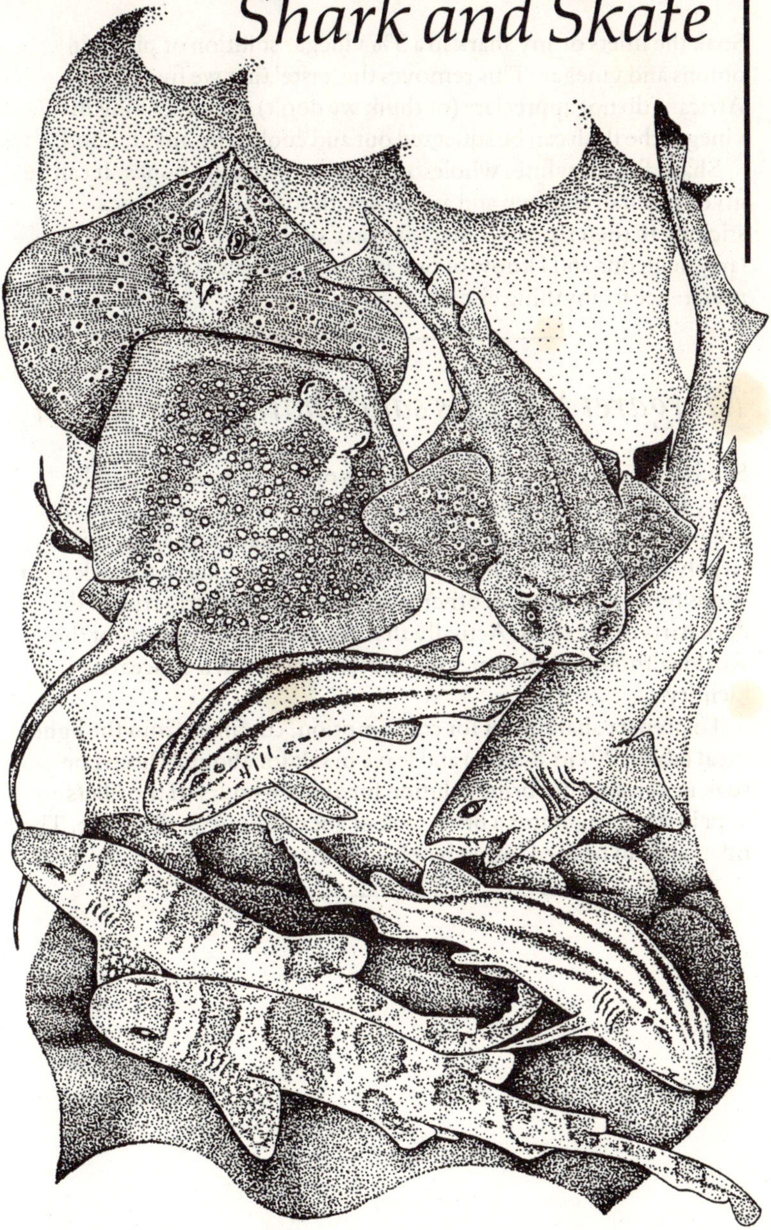

## *Margaret Smith's 'shark treatment'*

Soak the fillets of any shark in a 5% vinegar solution or pickle in onions and vinegar. This removes the 'taste' that we fussy South Africans do not appreciate (or think we don't). After soaking in vinegar the flesh can be squeezed out and cooked by any method.

Shark flesh is white, wholesome, tasty and has no bones. It can be given to small children and animals without fear of the bones sticking in their throats. The 'ruggraat' of the sand shark is particularly delicious and I know of one angler at Knysna who eats it in preference to anything else.

---

## *Margaret Smith's sand shark preparation*

Sand sharks and almost all other kinds of small sharks are edible. Very large sharks and hammerhead sharks should not be eaten. A prejudice exists against the consumption of shark meat, but this is largely unfounded. Perhaps it is partly due to the fact that shark meat spoils more quickly than other fish. This is particularly true of the dark parts of the flesh. Actually shark is eaten in many parts of the world, although sometimes it is given names which conceal its identity.

To prepare shark meat for consumption, cut fillets from the light meat (about 20 cm × 2 cm) and leave in salt for about 24 hours or soak in a clean, cold brine solution for about 6 hours. If the fillets are kept in the refrigerator a brine soak of about 2 hours will suffice. The fillets should be used immediately or else be frozen.

---

# *Jean Pote's shark and avocado cocktail*

Skin shark, soak in vinegar, rinse and then either steam or boil until cooked. Flake and mix with seafood mayonnaise. Cut avocados in half lengthwise and sprinkle with a little lemon juice. Spoon shark mixture into avocados and sprinkle a little paprika over each serving. Decorate with prawns or parsley and serve as a starter.

---

# *Shark braai sosaties*

1 kg skinned shark steaks *or* fillets
½ cup lemon juice
½ cup salad oil *or* olive oil
2 T chopped parsley
1 t dry mustard
1 clove garlic, minced *or* pressed
½ t salt
¼ t pepper
about 15 bay leaves

Wipe fish with a damp cloth and cut into 3 cm cubes. In a bowl, combine lemon juice, oil, parsley, mustard, garlic, salt and pepper and stir until blended. Pour two thirds of lemon marinade over fish cubes making sure the fish is well coated, then cover and chill for 30 minutes. Pour remaining marinade into serving bowl to serve as a lemon sauce. Thread fish on skewers, alternating every piece with a bay leaf. Place on grid about 15 cm above moderately hot coals, and baste well with marinade. Turn once, basting frequently until fish is browned and flakes readily when prodded (allow 5–10 minutes on each side). Serve with lemon sauce.

---

# *Shark patties*

1 cup minced shark
4 potatoes
½ t pepper
1 T butter
1 egg

Wash, peel and cut potatoes into pieces. Cook fish and potatoes in boiling water for 20 minutes or until potatoes are soft. Drain, add butter and pepper, and mash finely with a fork. As soon as the mixture has cooled, add the egg, well beaten, and salt if necessary. Shape into patties and fry in deep oil. Serve with lemon wedges.

# *Fried fillets of shark*

1 kg shark
½ cup flour
2 eggs
breadcrumbs
few sprigs parsley
1 lemon
salt and pepper
oil for frying
tomato sauce

Cut the fish into fillets, add salt and pepper generously, and roll in flour. Dip in beaten egg and roll in bread crumbs. Fry in deep oil until brown. Drain and serve garnished with parsley and slices of lemon, with tomato sauce on the side.

# *Shark marseillaise*

1 kg shark, cubed
2 large onions
2 T oil
4 tomatoes
1 clove garlic
salt and pepper to taste
125 m*l* water *or* fish stock
French bread, sliced and toasted

Chop the onions finely and fry in oil. Cut the tomatoes into small pieces and add garlic, salt, pepper and water or fish stock. Put the fish cubes in the mixture, and boil fast for 15–20 minutes. Keep the pot covered tightly. Remove the fish and place on slices of French bread. Reduce the liquid by boiling, adjust the seasoning, and pour over the fish.

---

# *Boiled skate parisienne*

1,5–2 kg skate *or* ray wings
⅔ cup vinegar
2 cups cold water
1 T salt
2 small carrots, sliced thinly
2 medium-sized onions, sliced thinly
½ t pepper
2 cloves
2 bay leaves

Put all the ingredients except the fish in a pot, bring to the boil and simmer for 30 minutes. Strain the liquid and simmer the fish in it for 20–30 minutes or until tender. Remove the skin and serve at once with caper sauce.

### Caper sauce

3 T butter
1½ T flour
⅔ cup hot water
½ t lemon juice
½ cup drained capers

Melt half the butter and blend with the flour. Gradually stir in the hot water. Boil for 5 minutes and stir in the lemon juice and the rest of the butter, followed by the capers.

---

## *Skate wings à la française*

1 kg skate wings, washed and diced
4 medium-sized onions, sliced
2 cloves garlic, chopped
37,5 m*l* butter
25 m*l* flour
500 m*l* milk
125 m*l* chicken stock
1 bay leaf
2,5 m*l* salt
black pepper
60 m*l* Gruyère cheese

Sauté the onions and garlic in butter, stir in the flour and stock, and slowly add the milk. Simmer until creamy. Add the seasoning and the cheese. Poach the skate wings in the sauce for 10 minutes. Put in a serving dish and serve with boiled potatoes. (Serves 4–6.)

---

# *Ray or skate with capers*

750 g skate *or* ray wings
2 t wine vinegar
salt
freshly ground white pepper
3–4 T capers
75 g butter

Cut the wings into 4 or 8 pieces. Pour enough water into an enamelled or stainless steel saucepan to cover the skate pieces, add 1 T of the vinegar and some salt, and bring to the boil. Lower the heat, add the fish and poach at a very gentle heat for about 15 minutes, until done. Remove and drain the fish and transfer to a warmed serving dish. If you wish, remove the skin and cut away the cartilage that runs along the inner side of each wing. Pepper the fish and strew with capers. Heat the butter in a pan until it is a light golden-brown, then pour it over the fish and capers. Heat the remaining vinegar in the same pan and sprinkle it over the fish. (Serves 4.)

---

# *Fried skate*

1–2 kg skate *or* ray wings
vinegar court bouillon
nutmeg
flour seasoned with salt
pepper
4 T butter

Cut skate into serving pieces. Simmer the fish in 250 m*l* court bouillon for 15 minutes. Dry thoroughly and roll the pieces in the seasoned flour. Fry in hot butter until brown. Take care when turning as they may fall apart.

## Vinegar court bouillon

1 cup vinegar
1,5 *l* cold water
1 T salt
2 small carrots, sliced
1 large onion, sliced
2 bay leaves
2 cloves
12 peppercorns
1 t thyme

Combine all the ingredients and simmer for 30 minutes. Strain.

# Ocean Favourites

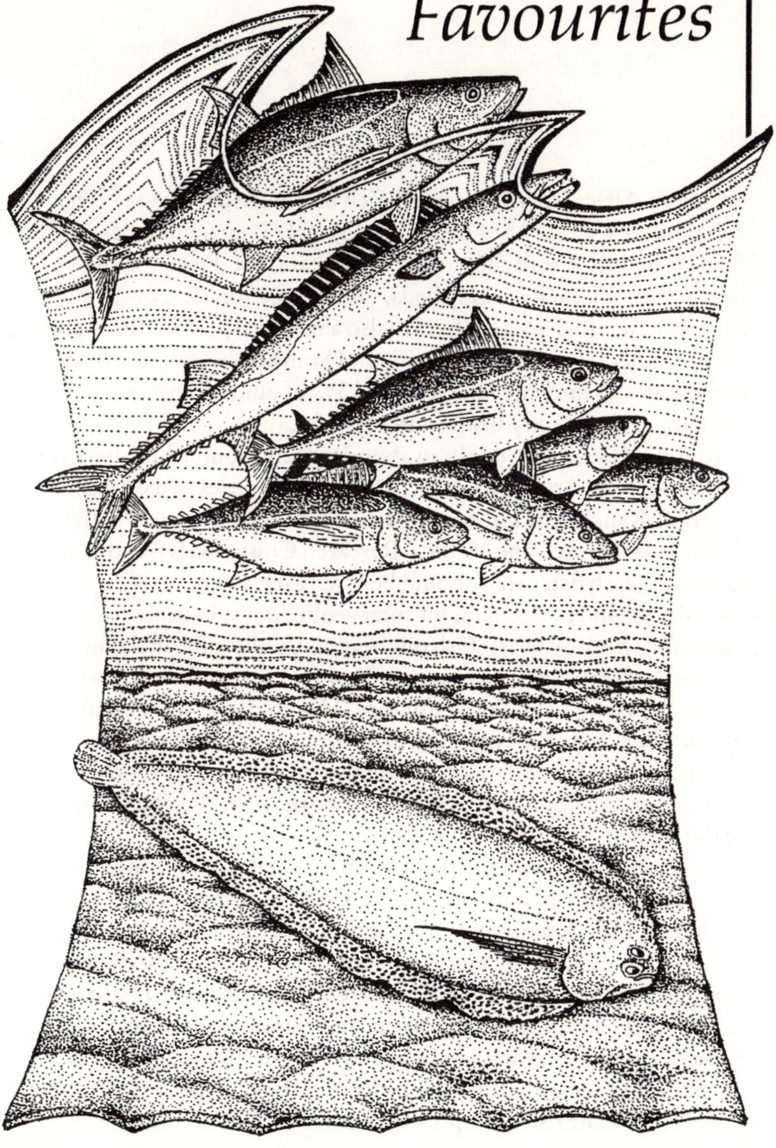

# INTRODUCTION

Southern Africa has an extremely rich marine fish heritage – over 2 200 species, about 15% of the world total. The richness of our fauna is mainly due to the variety of habitats: coral reefs, sandy beaches, rocky shelves, kelp beds, mud flats, estuaries, open water and ocean depths. South Africa is also a meeting place of three great oceans – the Atlantic, Indo-Pacific and Southern – each of which contributes to the fish fauna. About 13% of our fish species are endemic (known only from southern Africa).

Marine fishes are not only an abundant food source but also a very healthy one. Research has shown that communities that eat a great deal of fish such as Eskimos and Japanese fisherfolk, rarely suffer from heart disease. The Medical Research Council advises that one or two oil-rich fish meals should be eaten each week, and fish oil capsules should be taken on the intervening days, in order to maintain a healthy heart. Fish oils may also reduce the incidence of tumours of the colon and breast and have an anti-inflammatory action which helps arthritis sufferers. The lipid in fish is also an important component of the diet for brain and nerve tissue development. A fish diet is particularly important for pregnant mothers as fish lipids are extremely important for the development of the foetus.

Research in England has revealed that the fatty acids found commonly in oil-rich marine fish may make up more than 60% of the brain. This finding supports the idea that mankind evolved in close harmony with the marine environment, building settlements along estuaries and lagoons. Because their diet depended both on the rich marine food chain as well as on land-based foods, our ancestors had a distinct nutritional advantage over other mammals and so developed the huge brains which characterise our species.

So, eat marine fish to improve your health and intelligence!

# *Fish batter*

1 cup sifted cake flour
½ t salt
1 egg, separated
1 T oil
1 cup milk
2 T lemon juice

Beat flour, egg yolk and milk well together. Add lemon juice and oil. Lastly fold egg white into mixture. Dip fish in batter and fry in hot oil.

---

# *Carolynn's batter for deep-frying*

125 g flour
salt and pepper
1 T olive oil
150 m*l* beer *or* water
2 egg whites

Sift the flour into a bowl and season with salt and pepper. Make a well in the centre. Add the oil and gradually whisk in the beer or water, working from the centre outwards. Whisk for only as long as it takes to produce a smooth batter; do not overwork the mixture. Leave the batter to rest for about 1 hour at room temperature, or it will shrink away from the fish pieces and provide an uneven coating. Beat the egg whites until they form soft peaks, and fold them gently into the batter mixture just before using.

---

# HAKE

## *Humph's hake Godetia*

Allow one medium-sized hake (*Merluccius capensis*) per person. Scaled, gutted and filleted fish should be collected from the cod-end or fish markets as soon as practicable. Coat liberally with batter and deep fry. Serve on thick, very hot, china plates, with about an equal volume of thick, deep-fried potato chips. Garnish liberally with home-made pickled onions (about the size of a ping-pong ball). I know of no wine sufficiently indelicate to complement this dish, only strong black tea will suffice. Guests will particularly appreciate the piquancy of Hake Godetia if they are tired, cold, wet, have been waiting for some time and the table is arranged so as to pitch gently back and forth, occasionally rolling to one side or the other. Hake Godetia is named, with fond memories, for I & J's side-trawler 'Godetia' and her cooks (1947–1948).

---

## *Jean Pote's hake and tomato bredie*

500 g frozen hake, cut into serving portions
4 large tomatoes
1 large onion and a small piece of garlic
1 T vinegar *or* lemon juice
2 t sugar
1 finely chopped chilli (optional)
chopped parsley

Fry onion and garlic in oil. Add well floured and seasoned hake. Cook with lid on for 10 minutes, shaking pan. Add all the other ingredients except parsley and cook for another 10 minutes. Sprinkle with parsley and serve.

---

# *Hake tetrazzine*

500 g hake, steamed and flaked
250 g cheddar cheese, grated
250 g shell noodles, cooked
12,5 m*l* margarine
2 onions, chopped
1 green pepper, seeded and chopped
4 rashers bacon, chopped
1 tin cream of mushroom soup (410 g)
25 m*l* dry sherry
salt, pepper and paprika to taste

Fry onions, green pepper and bacon in margarine. In an ovenproof dish layer the onion mixture, noodles and seasoned fish. Mix soup, sherry and half the cheese and pour over. Sprinkle with remaining cheese and bake at 180° C for 30 minutes. Serve with a green salad and garlic bread. (Serves 4.)

# *Len Compagno's bacala Espanol*

250 g hake fillets, cut into small slivers or chunks
150–500 g white wheat flour
handful of fresh parsley
olive oil
3 cloves garlic
3 large eggs

Put hake in a large bowl with 500 m*l* water. Add egg and stir until well mixed with hake and water. Chop the parsley and garlic and add to the mix. Stir in enough flour to thicken mix to the consistency of pancake batter. Cover the bottom of a large frying pan with olive oil and heat until smoking, then add mix with a large spoon or ladle. Up to three cakes can be fried at once. Fry on one side until edges are crispy and mix has hardened, then flip cake and briefly fry on the other side. Replenish oil when necessary. (Serves 2–4.)

# *Party hake in aspic*

900 g hake fillets
75 m*l* salt
6 peppercorns
15 m*l* capers
10 m*l* lemon juice
1 bay leaf
1 hard-boiled egg

## Aspic

25 m*l* gelatine
1 chicken stock cube
15 m*l* lemon juice
30 m*l* dry white wine *or* sherry
600 m*l* water
1 small onion
3 m*l* pickling spice
muslin and ring mould

Poach the hake with salt, lemon juice, bay leaf and peppercorns. Cool, then flake and remove bones and skin. To make aspic, soak the gelatine in 60 m*l* water. Simmer the remaining water with the stock cube, onion, wine and pickling spice for 10 minutes, then stir in gelatine until dissolved. Add lemon juice and strain through a double layer of muslin. Pour a little into the base of a 1 litre ring mould and chill to set. Arrange eggs and capers in mould and carefully pour on a little more aspic liquid. Leave to cool. Fork the flaked fish into remaining aspic, pour into mould and refrigerate until set. Serve with a green salad and seafood dressing.

# *Crumbed baked fish*

750 g hake *or* kob, filleted and cut into serving portions
125 m*l* mayonnaise
250 m*l* cornflakes, crushed
salt and pepper

Season fish and coat both sides with mayonnaise, then roll in cornflakes. Refrigerate for 1 hour. Put a little oil in a baking dish and place fillets in hot oil, turning after baking for a few minutes. Bake at 180° C for about 20 minutes. Serve with mashed potatoes and peas.

# *Cape fish bredie*

1–1,5 kg hake
1,5 kg cabbage, shredded and parboiled
1 cup shredded celery (optional)
1 onion, sliced
4–6 potatoes, diced
salt and pepper to taste
little cooking oil

Heat oil in pan. Add onions and brown. Add cabbage, celery and seasoning. If 'wet' fish is used, add a little water; if frozen fish is used, no water should be added. Put fish and potatoes in saucepan, cover and simmer gently for about 30 minutes or until fish and potatoes are done. (Serves 8–10.)

---

# *Sweetcorn and fish pie*

500 g cooked fish of any kind, cut into bite-sized pieces
250 g corn (mealies), cooked *or* whole kernel tinned corn, drained
lemon juice
3 eggs
salt and pepper
breadcrumbs *or* cornflakes
2 T grated onion
1 T chopped parsley
butter

Mix the fish thoroughly with the onion, parsley and lemon juice. Beat the eggs well and add half to the fish and the remainder to the cut corn. Put half of corn mixture in a greased pie dish, place fish on top and pour remainder of corn over the fish. Sprinkle with breadcrumbs and dot with butter. Bake at 180° C until golden brown. Serve with vegetables or salad. (Serves 6.)

---

# *Fish and mushroom pie*

500 g hake fillets, cooked and flaked
1 cup green peas, cooked
1 small tin mushrooms (285 g)
1 T butter
1 cup milk
pastry
salt and pepper to taste
1 T flour

Melt butter. Stir in flour. Add milk, stirring constantly until sauce is smooth and thick. Add fish, peas and mushrooms. Season. Turn into a greased baking dish, top with rounds of pastry and bake at 200° C for 15 minutes. (Serves 4.)

---

# *Baked crumbed fish*

500 g frozen hake fillets
1 T cooking oil
1 t salt
½ t nutmeg
2 large tomatoes, sliced
2 T chopped chives
125 g chopped mushrooms
1 T sherry
1 onion, thinly sliced
¼ t pepper
⅛ t cayenne pepper
½ cup buttered crumbs

Oil a shallow baking dish, large enough to hold the fish in a single layer. Spread evenly with the sliced onion, then put the fish over the onion. Sprinkle fish with the salt, pepper, nutmeg and cayenne. Top with sliced tomato and sprinkle with chives. Scatter mushrooms over the fish, bake at 200° C for 15 minutes, then top with the buttered crumbs. Sprinkle with sherry. Continue baking until the crumbs are brown and the fish is done. (Serves 4–6.)

---

# *Hake fillets florentine*

500 g frozen hake fillets
2 bunches fresh spinach (300 g)
1 level T flour
2 T butter
½ t salt
½ t pepper
½ cup grated cheese
1½ cups milk

Cook spinach in 1 cup of water for 7 minutes. Drain and chop finely. Put in baking dish. Melt the butter and blend in flour and seasoning. Add milk and cook until thickened, stirring constantly. Add cheese and continue heating until melted. Pour sauce over spinach. Put fish fillets on top and bake at 190° C for 30 minutes. (Serves 8.)

---

# *Deep fried fish pasties*

1 cup cold, flaked hake
180 g plain flour
90 g margarine
2 t finely chopped onion
oil for frying
salt and pepper
iced water
tomato sauce

Sift flour with a little salt into a bowl and rub in margarine until mixture resembles fine breadcrumbs. Mix to a smooth dough with iced water, roll out and divide into 4 rounds. Mix fish, seasoning, tomato sauce and onion and divide between pastry rounds. Fold over and seal well. Fry in deep oil until golden brown. Drain on absorbent paper. (Serves 4.)

---

## *Fish cakes*

3 cups cooked, flaked hake
2 cups mashed potatoes
1 egg
1 T butter *or* margarine
1 T chopped parsley
½ t nutmeg
salt and pepper to taste
breadcrumbs

Cream butter. Add remaining ingredients, except breadcrumbs, and mix well. Form into fish cakes. Roll in breadcrumbs. Fry in oil until golden brown on both sides. (Serves 6–8.)

---

## *Soufflé of hake*

500 g cooked, flaked hake
3 eggs, separated
1 t salt
1 cup white sauce of medium thickness

Combine fish, salt, beaten egg yolks, white sauce and finally fold in stiffly beaten egg whites. Pour into a greased baking dish and bake in a moderate oven at 180° C for 25 minutes. Alternatively, put mixture in baking shells, top with grated cheese and bake until brown. (Serves 4.)

---

# *Fish fritters*

500 g frozen hake fillets
dash pepper
1 T finely chopped parsley
3 eggs, separated
½ t salt
dash minced garlic *or* garlic salt

Cook fish, remove skin and mash. Beat egg yolks until light and thick. Then add flour, salt, pepper, garlic or garlic salt, parsley and fish. Fold in stiffly beaten egg whites. Drop spoonfuls into deep hot oil and fry until brown. (Serves 6.)

# *Nutty fish dish*

500 g cooked, flaked hake
1 cup finely chopped nuts
cracker crumbs
butter
2 cups white sauce of medium thickness
2 hard-boiled eggs, finely chopped

Combine fish, white sauce, nuts and egg. Pour into a greased casserole. Cover with cracker crumbs and dot with butter. Bake at 180° C for 25 minutes. (Serves 6.)

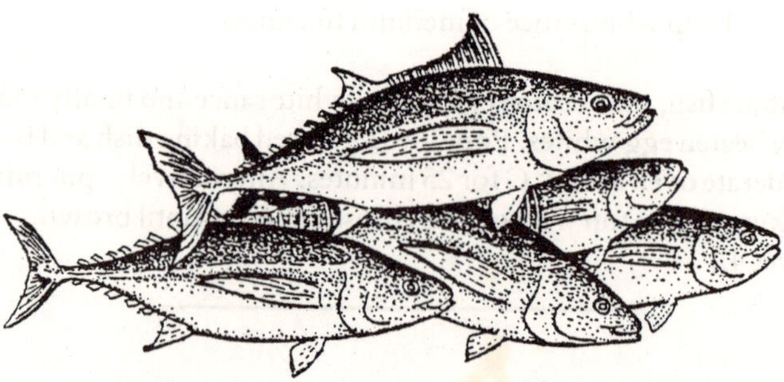

# *Fish rarebit*

500 g frozen hake fillets, cooked and flaked
1 cup tomato purée
2 t prepared mustard
2 eggs, beaten
125 g tangy cheese
1 t salt
1 T Worcestershire sauce
1 cup evaporated milk

Melt cheese over boiling water. Blend in purée and seasoning gradually. Add eggs to milk and stir slowly into cheese mixture. Add fish and heat for 5 minutes. Serve on toast. (Serves 6.)

# *Party fried fish with bacon*

5,5 kg hake fillets
flour, egg and breadcrumbs
50 slices of tomato, fresh *or* grilled
2 kg unsmoked streaky bacon, cut wafer thin

Coat the fillets with flour, egg and breadcrumbs in the usual manner and fry in hot oil. Drain well. Fry or grill the bacon to a moderate degree of crispness. Grill the tomato slices if desired. To serve, lay two slices of bacon in the form of an X on each portion of fish. Place a slice of tomato in the centre.

# HADDOCK

## *Liz's haddock pâté*

150 g smoked haddock, cooked (don't allow it to cool)
80 g melted butter
1 t lemon juice
¼ t hot curry powder
1 T chopped chives
pinch black pepper
pinch chilli powder

Blend the first 3 ingredients and mix in the rest with a spoon. Firm in the deepfreeze. Defrost for at least 1 hour before serving.

---

## *Graham Bell-Cross's smoked haddock quiche*

½ smoked haddock
shortcrust pastry
4 eggs
300 m*l* milk *or* 150 m*l* milk and 150 m*l* water
salt and freshly ground black pepper
fresh nutmeg
butter

Soak haddock in water for 1 hour. Drain, put in a small saucepan and cover with equal amounts of water and milk. Bring to a fast boil. Remove from heat and allow to stand for 15 minutes. Drain fish, reserving stock. Remove skin and bone from fish and flake. Whisk eggs with 300 m*l* milk (*or* 150 m*l* milk and 150 m*l* water) and 125 m*l* of reserved stock. When well mixed, season to taste with salt, freshly ground black pepper and fresh nutmeg. Line a large pastry dish with pastry, prick bottom of case and bake blind at 220° C for 10–15 minutes, or just long enough to set the crusts without browning them. Allow to cool. Arrange flaked, cooked haddock in bottom of pastry case and fill with the custard mixture. Dot with butter. Bake at 140° C for 30–40 minutes and serve immediately.

---

# *Savoury fish pie*

1½ cups cooked haddock *or* snoek
1 cup asparagus pieces (retain liquid for sauce)
1 cup grated cheese
1 egg, well beaten
1 T chopped parsley
salt and pepper
butter
shortcrust pastry

## White sauce

3 T flour
2 cups milk
3 T butter

Make white sauce with flour, milk, butter and asparagus liquid. Add remaining ingredients and mix well. Season to taste. Pour into a buttered shortcrust pastry shell and bake at 140° C for 35–40 minutes. Tomato slices can be placed on top of the mixture — it adds more flavour and looks very attractive.

---

# *Haddock and egg croquettes*

*This dish may be prepared a day in advance and kept in the refrigerator.*

2 cups cooked, smoked haddock, skinned and flaked
1½ cup butter
1 cup milk
1 t onion, finely chopped
oil for frying
1½ T flour
salt, pepper and nutmeg to taste
breadcrumbs
2 hard-boiled eggs, finely chopped
1 raw egg

Make a thick cream sauce by melting butter, adding flour and then milk and bringing it to the boil, stirring constantly. Add seasoning, fish, hard-boiled eggs and onion. Shape mixture into balls with floured hands, dip alternately into egg and then breadcrumbs, and fry in hot oil until golden. Serve with lemon wedges or tomato sauce. (Serves 4–6.)

---

# *Creamed haddock on buttered toast*

½ fillet of smoked haddock, skinned and flaked
2 t flour
2 hard-boiled eggs, chopped
2 t butter
½ cup milk
pepper

Poach haddock, changing water several times to remove salt as required. Make a cream sauce from butter, flour and milk, season lightly with pepper and add haddock and eggs. Pile onto hot buttered toast with crusts removed. Decorate with parsley and lemon. (Serves 4.)

---

# *Kedgeree*

375 g poached, smoked haddock
2 hard-boiled eggs, chopped
375 m*l* cooked long-grain rice
60 g butter
300 m*l* milk
1 egg, beaten
15 m*l* freshly chopped parsley
salt and pepper

Skin, bone and flake fish. Add to rice together with eggs. Melt butter in a large pan. Mix milk and beaten eggs. Add rice mixture to melted butter and stir in egg mixture. Cook very gently, stirring constantly, for about 2 minutes until heated through and well combined. Season lightly, sprinkle with chopped parsley and serve immediately.

---

# *Poached haddock in egg sauce*

1,5 kg haddock
250 m*l* milk
45 m*l* freshly chopped spring onions
2 m*l* salt
freshly ground black pepper
10 m*l* prepared mustard
1 large egg
10 m*l* chopped parsley

Wash haddock and pat dry between paper towels. Place whole in a pan and cover with milk. Poach for 15–18 minutes. Remove and cover with tinfoil to keep hot. Add spring onions to milk, season with salt and pepper and stir in mustard. Beat egg lightly and add to milk mixture. Then add 15 m*l* warm milk. Blend. Add to milk in pan, stirring constantly. Mix in parsley and serve immediately over hot fish.

# SALMON

## Salmon starter

1 tin red salmon (500 g)
100 g butter, melted
half a lemon
seasoning
parsley

Strain and carefully remove bone and black bits from salmon. Add juice of lemon, butter and a good pinch of chopped parsley. Mash well and serve chilled in egg cups. Top with a sprig of parsley.

---

## Salmon roll

### Pastry

2 cups flour
4 t baking powder
½ t salt
½ cup margarine
1 egg
1 cup milk

### Filling

1¾ cup salmon
¼ cup milk
2 chopped onions
¼ cup chutney
salt and pepper to taste
lemon juice to taste

Roll pastry (3 mm thick) into a rectangular shape. Spread mixture on top and roll up. Bake for 40 minutes at 180° C. Serve warm with a cheese sauce.

---

# *Salmon mould*

1 large tin salmon (500 g)
2 cups hot water
2 hard-boiled eggs, sliced
1 T gelatine
juice of ½ lemon
salt and pepper to taste
lettuce, tomato and lemon to garnish

Put salmon in a dish and flake with a fork. Soak gelatine in a little cold water and then dissolve in hot water. Add lemon juice, salt and pepper. Add this liquid to the salmon and mix well. Place egg slices in the bottom of a mould and gently pour in the mixture. Allow to set. Garnish with shredded lettuce and slices of tomato and lemon.

---

# *Salmon mousse*

2 × 250 g tins salmon
½ cup whipped cream
seasoning
½ cup salad cream
juice of 1 lemon
1 T gelatine
1 beaten egg white
1 cup stock (juice from fish with water added to make 1 cup)

Soak gelatine in ¼ cup cold water, then dissolve in heated fish stock. Flake salmon with fork, and add other ingredients. Lastly fold in egg white. Pour into damp mould and set in refrigerator. (Serves 4–6.)

---

# *Cold salmon soufflé*

2 × tins salmon (250 g)
chopped parsley
1 level T gelatine
4 tomatoes, puréed
4 egg whites, stiffly beaten

Mash salmon finely, add parsley, tomato purée and season to taste. Stir in gelatine dissolved in ¼ cup hot water. Fold in egg whites. Place in damp mould and set in refrigerator.

---

# *Salmon pie*

200 g shortcrust pastry
beaten egg or milk for glazing

## Filling

225 g salmon, flaked
65 g margarine
40 g flour
150 ml milk
50 ml cream
100 g sliced mushrooms
1 hard-boiled egg, chopped
3 t chopped parsley
salt and pepper
pinch nutmeg
1 tomato, skinned and sliced
100 g grated cheddar cheese

Roll out half the pastry and line a 20 cm pie dish. Heat 40 ml margarine, flour and milk in a saucepan. Bring to the boil, stirring constantly. Cook for 3 minutes until thickened, stirring all the time. Remove from heat and stir in cream. Cool. Melt the remaining margarine and sauté the mushrooms. Add the salmon and hard-boiled egg. Mix all this with the cold white sauce, parsley, seasoning and nutmeg. Spoon half the mixture into the base. Cover with a layer of tomatoes and sprinkle with the cheese. Spoon the remaining filling and roll out the remaining pastry and cover. Trim and decorate with a fork. Glaze with beaten egg or milk and bake at 180° C for 45 minutes.

# SNOEK

## *Phil Heemstra's Silwood snoek*

400 g smoked snoek, skinned, boned and flaked
juice of 1 lemon
freshly ground black pepper
125 m*l* cream
1 T freshly chopped parsley
pinch salt
60 g butter, melted

Combine all the ingredients, except salt and lemon juice, mixing well with a fork. Taste and add salt if necessary. Put one third of the mixture at a time in a blender, slowly adding the lemon juice and work until smooth. Put in a bowl and smooth the top with a spatula. Garnish with lemon slices, sprigs of parsley and black olives.

---

## *Savoury snoek tart*

500 g snoek
400 g shortcrust pastry
125 g mushrooms
180 g butter
100 m*l* cream
100 m*l* white wine
salt and pepper
2 egg whites, beaten
nutmeg

### Lemon Butter Sauce

50 m*l* melted butter
5 m*l* tarragon
5 m*l* chopped parsley
5 m*l* chopped chives
juice of half a lemon

Skin and bone the snoek; flake half and reserve. Pound the other half to a smooth paste in a mortar with mushrooms, butter, cream and wine. Season. Fold in egg whites. Line a pie dish with pastry. Half fill with snoek paste, cover with a thin layer of reserved snoek and season. Add remainder of snoek paste and cover with a second layer of pastry. Make a hole in centre of pastry crust and bake at 200° C for 30–40 minutes. Combine ingredients for lemon butter sauce and, just before serving, pour through hole in the top of crust.

---

## *Smoorsnoek*

1 kg uncooked smoked snoek
2 large onions, sliced
3 potatoes, sliced
1 T shortening (margarine, butter, oil *or* cooking fat)

Soak snoek for a few hours in enough water to cover. Boil until done. Drain, flake and remove bones. Boil onion and strain. Now brown in margarine or cooking fat. Add potatoes and a little water. When potatoes are cooked, add flaked fish and simmer slowly. Add a pinch of pepper. Stir well before serving. Serve with rice, peas and a green salad.

---

# *Snoek pie*

50 m*l* smoked snoek, flaked
30 m*l* flour
30 m*l* butter
500 m*l* milk
4 eggs, beaten
17,5 m*l* chopped onion
17,5 m*l* chopped parsley
10 m*l* Worcestershire sauce
250 m*l* cheddar cheese, grated
pinch cayenne pepper
1,5 m*l* mustard
grated nutmeg
salt

Make a white sauce with the butter, flour and milk. Add the eggs and then the cheese, chopped onion, parsley, snoek and seasoning. Turn into a greased ovenproof dish and bake for about 20 minutes at 200° C. Lower heat to 180° C after first 10 minutes.

---

# *Braised snoek*

500 g salted snoek
2 onions, sliced
3 boiled potatoes, cubed
3 ripe tomatoes, skinned and sliced
3 m*l* sugar
salt and pepper
butter
nutmeg

Soak snoek overnight in cold water. Drain and boil in fresh water until done. Flake fish with fork and remove bones. Brown onion slightly. Add tomatoes and sprinkle with sugar. Fry tomato and onion mixture for a few minutes, then add snoek, potatoes, a pat of butter, grated nutmeg, salt and pepper. Braise at 140° C. Add water if necessary and stir regularly. Serve with rice.

---

# *Snoek vinaigrette*

500 g smoked snoek, flaked
110 m*l* cream
10 m*l* chopped chives
10 m*l* chopped celery
125 m*l* capers
1 hard-boiled egg, chopped
10 m*l* vinaigrette sauce

Combine all the ingredients loosely and sprinkle with the sauce. Serve on a bed of lettuce with 2 lemon quarters.

## Vinaigrette sauce

1 clove garlic, crushed
5 m*l* sugar
5 m*l* mustard
chopped chervil, tarragon, parsley and chives
salt and black pepper
12,5 m*l* onion, finely chopped
125 m*l* oil
25 m*l* wine vinegar
25 m*l* tarragon vinegar

Chop herbs finely and blend all ingredients to make a tangy sauce.

---

# *Snoek and bean supper*

200 g smoked snoek, flaked
1 tin butter beans, drained (200 g)
4 T salad oil
1 clove garlic, crushed
3 onions, chopped
1 t salt
1 t marjoram
1 t sugar
1 t paprika
black pepper to taste
1 t lemon juice
1 tin tomatoes (100 g)
125 m*l* chicken stock

Heat oil in a frying pan and add garlic and onion. Sauté until tender. Add remaining ingredients and simmer for 5–10 minutes. (Serves 4.)

# KINGKLIP

## *Tessa Davies' kingklip bake*

450 g kingklip
90 g butter
15 m*l* grated onion
250 g mushrooms
15 m*l* chopped parsley
60 g butter
60 g flour
5 m*l* dry mustard
salt and pepper
celery salt
pinch cayenne pepper
onion salt
300 m*l* milk
250 m*l* grated cheese
juice and rind of ½ lemon
30 m*l* sherry
buttered crumbs

Steam fish for about 20–30 minutes. Set aside. Melt 90 g butter in frying pan and add onion, mushrooms and parsley. Fry gently. Remove and keep warm. Melt 60 g butter in the pan, add flour and all the seasoning. Stir over heat, gradually adding milk until the sauce thickens. Bring to the boil. Remove from the heat and stir in cheese, lemon juice, rind and sherry. Flake fish and add to the hot sauce. Add the cooked mushrooms and mix well. Pour into a greased ovenproof dish, cover with buttered crumbs and bake at 180° C for about 45 minutes.

# *Thora McGlouglin's baked fish in mustard cream*

500 g kingklip *or* hake *or* other suitable fish for baking
1 large onion, sliced
1 cup cream
10 m*l* mustard

Arrange pieces of fish in baking dish over onion. Mix cream and mustard and pour over fish. Bake at 180° C until fish is cooked.

---

# *Kingklip bouillabaisse*

400 g kingklip fillets, cooked and cut into large cubes
25 m*l* oil
1 leek, sliced
1 large onion, sliced
1 clove garlic, crushed
2 large carrots, thinly sliced
1 small bay leaf
10 m*l* turmeric
750 m*l* fish *or* chicken stock
250 m*l* dry white wine
4 slices of bread for croutons

Heat oil and fry onions and garlic until transparent, but not brown. Add carrots, leek, bay leaf and turmeric. Pour over wine and stock. Boil until carrots are soft. Add fish and simmer for another 15 minutes. Add salt and pepper to taste. Deep fry croutons and serve immediately. (Serves 4.)

---

# *Kingklip with angel sauce*

required number of kingklip fillets

## Sauce

300 g button mushrooms, sliced
250 m*l* cream
¼ t garlic salt
pinch salt
freshly ground black pepper

Fry or grill the fish. Combine the sauce ingredients and simmer for 15 minutes. Pour over fish and bake at 150° C for ½ hour.

---

# *Kingklip delight*

2 kg kingklip, steamed, cooled and flaked
1 cup sliced onions
500 g fresh mushrooms, sliced
25 g margarine
1 tin celery soup (410 g)
1 tin tomato soup (410 g)
½ t salt
1 cup sliced green peppers
¼ t thyme
¼ t chilli powder
¾ t tabasco

Sauté onions, peppers and mushrooms in melted margarine for about 10 minutes. Add seasoning, celery and tomato soup. Blend well and simmer for another 10 minutes. Remove from stove and add kingklip. Add more salt if necessary. Serve immediately with rice. Alternatively put rice in a mould, unmould on a hot platter and fill the centre with kingklip mixture. Arrange peas or Brussels sprouts on the outside. (Serves 6–8.)

---

# *Stuffed kingklip*

500 g frozen kingklip fillets
1 cup fresh breadcrumbs
1 egg
½ t mixed herbs
salt and pepper
a little chopped onion
grated cheese
1 cup thin white sauce

Make a stuffing with breadcrumbs, egg, herbs, onion, salt and pepper. Spread between fish fillets. Bake in oven at 200° C for 10 minutes. Remove from oven, cover with white sauce and bake for 20 minutes. Sprinkle grated cheese over and bake until cheese is golden brown.

# SOLE

## *Linda's sole with cream sauce*

4 soles
salt and pepper
a little cayenne pepper
300 m*l* cream
1 tin creamed mushrooms (285 g)
1 tin mushrooms (285 g)

Skin, wash and fillet the soles and divide each fillet into 2 pieces. Place in cold, salted water and gradually bring to the boil. Once the water has boiled, drain and mix the cream, creamed mushrooms, mushrooms and seasoning. Pour over the sole and simmer gently for 15 minutes. (Serves 4–5.)

## *Baked sole*

2 soles, washed, skinned and boned
salt and pepper
grated nutmeg to taste
egg white
breadcrumbs
butter
juice of 1 lemon

Divide each sole fillet into two pieces. Brush with egg white, sprinkle with breadcrumbs and seasoning, and place in a baking dish. Dot with butter and bake for 30 minutes. When the fillets are nearly done, sprinkle with lemon juice. Serve with sauce tartare. (Serves 4.)

# Fried sole

Beat an egg with a little lemon juice. Coat soles with seasoned flour, dip into the egg and then in breadcrumbs. Fry in hot oil until golden brown. Garnish with lemon slices.

---

# Curried baby sole

500 g baby soles
1 clove garlic
½ cup chopped onions
1 t strong curry powder
½ cup vinegar
½ cup tomato purée
salt and pepper
3 bay leaves
peppercorns
1 T apricot jam
1 cup water
2 T cornflour
1 T oil
flour
1 egg, beaten

Fry garlic and onion in oil. Mix curry, vinegar and tomato purée; add to fried onions and garlic. Add salt, pepper, bay leaves and peppercorns. Add apricot jam and water. Simmer for 30 minutes. Add cornflour mixed with a little water. Stir into curry mixture and cook for another 10 minutes. Remove from stove and cool completely. Coat soles with flour and beaten egg. Fry in hot oil. Put in dish, cool and pour over the curry sauce. This dish is at its best 48 hours after preparation.

---

# TUNA

## *Roy Bross's fish parcels*

*An unusual starter or supper snack.*

### Pastry
113 g butter
113 g cream cheese
227 g flour

Combine butter and cream cheese in mixing bowl and stir until blended. Using a fork, add flour and mix well. Knead dough and refrigerate for 1 hour.

### Filling
2 × 200 g tins tuna *or* middle-cut mackerel
5–10 m*l* lemon juice
½ t curry powder
1 T finely chopped parsley
salt and black pepper

Bone fish and mash with fork. Add lemon juice, curry powder, salt and pepper to taste. Mix in finely chopped parsley.

Roll dough 3–4 mm thick on floured surface. Cut pastry into oblongs 5 mm × 10 mm. Place a heaped teaspoon of filling at one end of each oblong. Turn other end over to produce the parcel (square) and seal. Place on a baking sheet and bake at 200° C for about 10 minutes until golden brown.

# *Tuna in green olive sauce*

6 fresh tuna steaks
flour
oil for frying
150 g chopped green olives and a few whole, for garnishing

## Marinade

2 T olive oil
1½ glass dry white wine
3 cloves garlic, minced
1 T vinegar
1 bay leaf
1 pinch dried thyme *or* two fresh sprigs
salt and freshly ground black pepper

Mix ingredients for marinade and pour over tuna. Refrigerate for 3 hours, turning once or twice. Remove tuna, coat with flour and fry in oil in a large pan until golden brown. Place in serving dish and keep warm. Add marinade to pan, stir well and boil until reduced by half. Remove from heat, strain the sauce, add the chopped olives and reheat, stirring. Pour over tuna, garnish with whole olives, serve immediately.

---

# *Vanessa's tuna and mushroom*

1 tin tuna, drained and flaked (200 g)
6 T grated cheese
20 large button mushrooms, chopped
½ cup milk
salt and pepper to taste, pinch of marjoram
cornflour, dissolved in 2 t water

Cook mushrooms in milk with salt, pepper and marjoram for about 10 minutes or until tender. Add cornflour to thicken. Blend in cheese, one teaspoon at a time, until melted. Stir in tuna. Leave to cook through and serve on rice, spaghetti or toast. (Serves 2.)

---

# *Tuna and vegetable hors d'oeuvre*

1 tin tuna, drained (200 g)
1 tin curried mixed vegetables (410 g)
4 T mayonnaise
2 t fruit chutney
pinch of freshly ground black pepper
2 hard-boiled eggs, finely chopped
chopped parsley

Flake tuna lightly and add curried vegetables. Mix in mayonnaise, chutney and black pepper. Spoon individual portions on to crisp lettuce leaves in scallop shells or in suitable glass dishes. Sprinkle with egg and parsley. Serve with thinly sliced brown bread. (Serves 6.)

---

# *Tuna pie*

## Crust

1 packet savoury biscuits, crushed
125 g butter

## Filling

1 tin tuna (200 g), drained and flaked
1 jar cheddar cheese spread (170 g)
2 eggs, lightly beaten
1 onion, chopped
1 green pepper, chopped
5 m*l* mustard

Melt butter and mix with biscuit crumbs. Press into a greased pie dish, reserving a little to sprinkle on top of pie. Mix all the remaining ingredients and pour into the prepared crust. Sprinkle the reserved crumbs on top and bake at 170° C for 30 minutes. Serve hot or cold.

---

# Freshwater Originals

# EEL

Freshwater eels are arguably one of the most fascinating, elusive and resilient families of fish. They spawn in the Indian Ocean, probably at depths in excess of 1,5 km, in an area northeast of Madagascar. The small larvae drift southwards in ocean currents before changing into young eels that enter the rivers along the east and south coasts of the continent. Some of these elvers swim all the way into the mountainous headwaters of our rivers, a steep and often perilous journey, while others reside lower down in the river systems. After living in the rivers for up to 20 years, they undertake the long journey back to their ancestral spawning grounds, navigating with all the skill of an experienced seafarer!

We have four species of freshwater eels; two mottled yellow-green species and two plain yellow-brown species, and all make delicious eating – do *not* be deceived by their appearance! I have found all the local species very tasty but the smaller yellow-brown ones top the list. Worldwide there are 16 species of freshwater eels and 60 000 tons of European and Japanese eels alone are eaten every year! Unfortunately they have never been bred successfully in captivity. So, because they are a relatively scarce delicacy, they command one of the highest prices of all fish meats – retail prices for imported smoked Dutch eel can exceed R50 per kilogram.

## *How to catch eel*

The rivers and dams of the eastern and southern regions of South Africa hold good stocks of eels which will be much cheaper to catch than to purchase. The best months for eel fishing are November to March and the best times are during and shortly after good rains in the evening and early morning, although good catches can be made throughout the day and especially at night. Earthworms, bird meat, fresh prawn, crab or fish meat are all suitable baits. But remember to obey the angling regulations!

## *Preparation*

An eel, once obtained, should be gutted fresh with a ventral slit from 3 cm behind the anus through the gut wall to the head. Take care to remove all of the kidney, which lies beneath the spine. Unless the fish is to be smoked or prepared as Kabayaki, its skin must be removed. To facilitate handling and preparation, the mucous coat can be removed by rubbing salt over the surface and wiping it clean afterwards. Make a cut through the skin just behind the head and ease the skin away from the meat for about 2 cm. Grip this fold of skin securely in one hand and the head with the other; then pull the skin over the body and off at the tail. With larger eels this may be difficult and I have found it advantageous to nail the head to a tree and pull the skin off by using a pair of pliers!

Martin Davies

# *Smoked eel*

Eel is normally hot-smoked and its naturally high oil content lends itself well to this preparation. The freshly gutted, unskinned eel can be sprinkled lightly with salt before being smoked or put in a saturated brine solution of 1 kg salt to 4 litres water for 10 minutes.

Smoking boxes, readily available commercially, or a 200 litre steel drum can be used. The eels are suspended by their heads from steel bars inside the drum. A nonresinous fine sawdust is best for smoking, oak being the most commonly used. Smoking in the sealed smoking box normally takes no longer than 15 minutes for a 1 kg eel. The meat should readily flake off the backbone when ready.

# *Fry and braai*

To fry, cut the skinned and lightly salted eel into small pieces of about 5 cm and immerse in hot vegetable oil until cooked. This will normally take no more than 2–4 minutes. The pieces can then be browned under the oven grill at medium heat for a few minutes.

To braai, place the skinned and lightly salted eel over coals that are not too hot. Turn only enough to prevent the meat from scorching. The meat becomes tender when being cooked and must be handled carefully in order to prevent it from breaking up. Larger eels of over 2 kg should preferably be left unskinned. Run a length of 10 mm round iron through the body from head to tail. The lightly salted and skinned eel should then be spit-braaied. The eel cooks in its own natural fat, which tends to drip out, and should be turned regularly. It is best to place it well above the coals, which should smoke a little as this adds a delicious mild smoked flavour to the meat.

---

# *Kabayaki*

This is a traditional Japanese gourmet dish prepared from the freshest eels. The gut and backbone are removed and the 'fillets' cut into several pieces, 10–12 cm long. These pieces are then skewered on thin strips of bamboo or fine metal rods are threaded through each fillet to keep the meat flat while being cooked.

Boiling water is poured over the segments on the skin side, then the flesh side, before the segments are lightly cooked over steam. They are then dipped in soy sauce and grilled over charcoal. The dipping and grilling should be repeated a few times until the fillets are brown and crisp. This provides a succulent delicacy that melts in the mouth.

---

# *Syrup-coated fried eel*

450 g fresh eel, cut into 5 cm pieces
250 m*l* water
25 g butter
1 t salt
3 t golden syrup

Sprinkle salt over eel pieces, place in frying pan, add water and bring to boil, simmering until water has evaporated. Add butter and fry until golden brown. Add the syrup in thin 'drools' over the eel pieces, turn and continue frying until they are crisp. Serve with pan juices poured over the eel and with baby potatoes and a green salad. (Serves 4.)

---

# *Eel tartare*

1 kg eel, skinned, washed and cut into pieces
1 carrot, chopped
1 onion, chopped
flour
½ cup of sherry (125 m*l*)
2 T vinegar
salt and pepper
nutmeg to taste
breadcrumbs
1 egg

Stir-fry carrot and onion for 5 minutes. Sprinkle with a little flour, add sherry, vinegar and seasoning. Boil for 30 minutes, adding a little water from time to time. Add eel and simmer until tender. Remove eel, allow to cool, coat with breadcrumbs and fry until brown.
Place in serving dish, pour over remaining sauce and serve hot. (Serves 5–6.)

---

# *Eels in green sauce*

1 kg prepared eel, cut into 50 mm lengths
125 g margarine
3 t each of chopped fresh parsley, mint and sage
salt and pepper
250 m*l* dry white wine
250 m*l* water
3 egg yolks blended in 125 m*l* water
juice of one lemon

In a frying pan, heat half the butter, then add eel segments, salt, pepper and herbs. Fry, stirring carefully, for 10 minutes. Stir in egg yolk mixture, remaining butter and lemon juice. Heat gently and spoon into serving dishes. Serve chilled with lemon slices to garnish. (Serves 6.)

---

# *Boiled eel*

4 small eels
a large bunch of parsley

Choose small eels for boiling; put in a stew-pan with the parsley and add sufficient water to cover. Simmer until tender. Remove eels, pour a little parsley and butter over and serve. (Serves 4.)

---

# CATFISH

Why eat catfish? Take one look at a catfish and in all likelihood you will never want to see another, let alone taste it. On the other hand, remove the head and skin and cut the fillets from the vertebral column, prepare it in any way you like to eat fish, garnish it with your favourite sauce, and serve it to your unsuspecting guests as 'Blue Waterberg Spike', or 'Blyde Wyn's Tinka', and an unprecedented success is guaranteed. The fillets are completely boneless and have a reddish hue. When steamed, fried or prepared in a microwave oven, the flesh is flaky with a good texture and an excellent taste. When smoked it is unparalleled, and if cold smoked and sliced thinly, it is reminiscent of smoked salmon and gammon. The taste and texture of this indigenous species is completely underrated and is certain to surprise many a gourmet in the future.

To skin the fish a cut is made through the skin around the body, just below the head. The skin is then held with a pair of pliers and can be pulled off in one piece in a single swift tailwards pull. Once skinned, the fish can be filleted in the same way as any other fish.

Catfish are now being cultured on several farms in the Transvaal Lowveld and Northern Cape. Already the demand for the fish is in excess of what the farmers can produce. When next you catch one of these fish in a river, dam or lake, we urge you for once to throw prejudice overboard and try it. The recipes in this cookery book are tried and tested ones, made available for publication by the fortunate few who have had the chance to taste this fish. Try them and you'll smile!

Tom Hecht

## *Ben's smoked catfish*

Clean 0,5–1 kg catfish, split lengthwise and keep in a saturated brine solution for 30 minutes. Dry, secure with wire fasteners on rods across an open-ended 210 *l* drum, cover with a moist grain bag after the fire has been prepared, and smoke for 3–5 hours.

*To prepare the fire:* Make a small fire or use coals from a braai fire, cover with sawdust (*not* from pine; saligna bluegum can be used, and wattle, acacia and oak are all excellent). Position drum over the smoking fire. By allowing a little air to pass between the bottom and the surface, a good smoke with a low temperature is ensured! Small catfish don't have to be split – the smoke penetrates the skin easily.

---

# *Crisp fried catfish*

6 smallish catfish, skinned, filleted and cut into pieces
1 t salt
¼ t pepper
2 cups self-raising flour in brown paper or plastic bag
125 m*l* oil, *or* more

Sprinkle catfish lightly with salt and pepper. Drop pieces into bag of flour and shake to coat each piece well. Use a deep pot half filled with cooking oil. Heat oil until just under 'smoking hot'. Place each piece into oil separately. Cook on high until catfish floats to top and is golden brown. Drain well and place on paper towel. Serve hot.

---

# *Dixie-baked catfish*

6 catfish of suitable size
½ cup French dressing
12 thin lemon slices
paprika

Brush inside and out with dressing. Cut 6 lemon slices in half. Place 2 halves in each body cavity. Place catfish in a greased baking dish, with a lemon slice on top of each. Brush top of catfish with remaining dressing. Sprinkle with paprika. Bake at 180° C for 30–35 minutes, or until catfish flakes easily when tested with a fork.

---

# *Catfish creole*

500 g skinned catfish fillets, cut into 2 cm pieces
⅓ cup margarine *or* cooking oil
⅓ cup flour
½ cup sliced celery
1 cup sliced green onion and tops
½ cup chopped green pepper
2 cloves garlic, finely minced
1 tin tomatoes (410 g), cut into small pieces
250 m*l* tomato sauce
1½ t salt
2 bay leaves
½ t thyme
¼ t pepper
1 T brown sugar
1 T lemon juice
1 t Worcestershire sauce
¼ t hot pepper sauce
¼ cup chopped parsley
cooked rice

Brown flour in margarine. Remove from heat and cool slightly. Add 75 m*l* water gradually and stir until blended. Add remaining ingredients except catfish and rice. Cover and simmer for 20 minutes or until vegetables are tender. Remove bay leaves. Add catfish and simmer for another 10 minutes or until fish flakes easily when tested with a fork. Serve over rice.

---

# *Hollandaise catfish*

2 kg catfish fillets
1 egg
1 cup milk
1 cup cornflour
1 tin asparagus (410 g)
170 g crabmeat

Season fillets with salt and pepper, dip in egg-milk mixture, dredge in cornflour and fry in oil at 190° C. While fillets are frying, heat asparagus in saucepan and crabmeat in another saucepan. When fillets are cooked, place 3–4 spears of asparagus on top of each fillet, then place 1/4 of the drained crabmeat on the asparagus. Slowly pour over the Hollandaise sauce and serve.

### Hollandaise sauce

3 egg yolks
1 T lemon juice
1 T melted butter
1/8 t cayenne pepper

Put egg yolks, lemon juice and cayenne pepper in blender and mix well. Pour hot melted butter into egg yolk mixture and beat vigorously.

---

## *Fried catfish*

1 kg skinned catfish fillets, cut into serving-sized portions
2 t salt
1/4 t pepper
2 eggs, beaten
2 T milk
1 1/2 cups cornflour
fat for frying
lemon wedges

Season fish with salt and pepper. Combine eggs and milk. Dip fish in egg mixture and roll in cornflour. Put fish in a single layer in hot fat in a pan. Fry at a moderate heat for 7–8 minutes. Turn carefully and fry for another 7–8 minutes or until fish is brown and flakes easily when tested with a fork. Drain on absorbent paper. Garnish with lemon wedges. (Serves 6.)

---

# Catfish baked in foil

4 catfish fillets
⅓ cup lemon juice
¼ cup diced celery
½ cup diced carrots
1 t pepper
2 t salt
¼ cup chopped green onion
½ cup melted margarine
2 lemons, thinly sliced
2 T chopped fresh parsley

Place each fillet on a piece of foil about 10 cm longer than the catfish. Melt margarine and add lemon juice, salt, pepper, carrots, celery, onion and parsley. Pour quarter of mixture over each fillet. Top with lemon slices. Bring edges of foil together and fold over several times. Twist one end to form a tail. Tuck the other end under to form a point, or nose of the catfish. Place on baking tray and bake at 180° C for 30 minutes or until catfish flakes easily. Garnish with fresh parsley.

---

# Baked catfish with rice

1 kg catfish fillets
½ t salt
dash pepper
2 cups cooked rice
2 T grated onion
½ t curry powder
6 thin lemon slices
¼ cup butter *or* margarine
chopped parsley

Cut fish into serving-sized portions and put in a greased baking dish of 320 × 230 × 50 mm. Sprinkle with salt and pepper. Combine rice, onion and curry powder; spread over fish. Top with lemon slices and dot with butter. Cover. Bake at 180° C for 25–30 minutes or until fish flakes easily when tested with a fork. Remove cover for the last few minutes of cooking to allow slight browning. Sprinkle with parsley.

---

# *Margaret Jackson's maan nezim nzedo*

*This catfish stew with vegetables comes from Upper Volta.*

750 g freshwater fish, preferably catfish, cut into 6 pieces
½ cup vegetable oil
2 tins tomato purée
1 onion, thinly sliced
¼ t red pepper
2 t salt
1 packet frozen okra pods cut in halves
3 carrots, cut into 10 mm slices
1 small head of cabbage, quartered and sliced
1 packet frozen green beans
1 cup rice

Combine oil, tomato purée, onion, pepper and salt in a 30 cm saucepan and bring to the boil. Add the catfish, okra, carrots, cabbage and beans. Cover, bring to the boil and cook over low heat for 5 minutes. Add rice and cook for another 25 minutes. Add water during cooking, if necessary. (Serves 6.)

---

# *Cheesy catfish*

1 kg small catfish fillets
1 egg, lightly beaten
½ t salt
½ t pepper
¼ cup melted butter
1 T milk
1 t paprika
¼ cup flour
½–1 cup Parmesan cheese

Heat oven to 180° C. Combine cheese, flour and seasoning. Dip catfish in combined egg and milk, then coat with cheese mixture. Put in baking dish. Pour butter over catfish and bake at 180° C for about 30 minutes until golden brown. (Serves 4.)

# *Catfish cakes*

400 g catfish mince
100 g white bread
50 g onion, finely chopped
2 t salt
¼ t pepper
2 eggs
¼ t chopped parsley
¼ t origanum

Mix mince, bread and chopped onion. Add salt, parsley, pepper and origanum and mix well. Add eggs and mix well. Roll mixture into cakes and fry in hot oil for 10 minutes. Drain and serve garnished with sauce tartare on a heated plate.

# *Smoked catfish pâté*

350 g smoked catfish, cut into strips
1 clove garlic, crushed
120 g onion, chopped
100 g butter
30 m*l* anchovy spread
60 m*l* cream
2,5 m*l* pepper

Melt butter and fry onion and garlic until transparent. Add fish and fry until tender. Add anchovy spread, cream, and pepper. Liquidise to a fine paste. Serve on bread or salt biscuits, or as a filling for tomatoes or boiled eggs.

---

# *Catfish with savoury filling*

1,4 kg catfish, washed and dried
10 m*l* salt
5 m*l* black pepper

Preheat oven to 180° C. Cut through the length of the fish and sprinkle inside and outside with salt and black pepper.

## Filling

50 g butter
150 g chopped onion
1 clove garlic, crushed
20 g chopped parsley
40 g chopped bacon
1 tin mushrooms (90 g)
1 egg, beaten
250 m*l* fresh white breadcrumbs
5 m*l* salt
2,5 m*l* black pepper
30 g flaked almonds
75 m*l* white wine

Melt butter and fry onion, bacon, garlic and mushrooms until tender. Add salt, black pepper, breadcrumbs, beaten egg, parsley, almonds and white wine. Fill fish with the mixture and tie with string.

Basting sauce
20 g butter
5 m*l* lemon juice

Melt the butter and lemon juice. Bake fish at 180° C for 35–40 minutes. Baste regularly with butter and lemon sauce. Serve hot. (Serves 8.)

---

## *Fried catfish with creamy almond sauce*

750 g catfish fillets, washed and dried
150 g butter
100 m*l* cake flour
10 m*l* salt
5 m*l* pepper
60 g flaked almonds
1 tin mushrooms (285 g)
30 m*l* lemon juice
45 m*l* chopped parsley
120 m*l* thick cream

Cut fish into serving-sized portions. Season with salt and pepper and coat with cake flour. Heat half the butter and fry fish at moderate heat until it is brown and flakes easily when tested with a fork. Melt the remaining butter and fry almonds until golden brown. Add mushrooms, lemon juice and parsley to almonds and simmer lightly until the mushrooms are tender. Add cream and stir well. Serve hot over catfish portions. (Serves 5.)

---

# *Fried catfish with buttered almonds*

750 g catfish fillets, washed and dried
130 g butter
50 m*l* oil
100 m*l* cake flour
10 m*l* salt
5 m*l* pepper
180 g flaked almonds
20 m*l* chopped parsley
10 m*l* Worcestershire sauce
10 m*l* lemon juice

Cut fish into serving-sized portions. Heat 80 g butter with oil. Coat fish portions with a mixture of cake flour, salt and pepper. Fry in melted butter until golden brown.

## Topping

Melt 50 g butter and add almonds. Mix parsley, Worcestershire sauce and lemon juice and add to almond mixture. Put fried catfish portions in serving dish and pour over almond sauce. (Serves 5.)

---

# *Baked sweet and sour catfish*

1 kg catfish fillets, washed and dried
70 g diced onion
6 lemon slices
5 m*l* salt
2 m*l* pepper
125 m*l* tomato sauce
30 m*l* oil
30 m*l* dry white wine
30 m*l* chopped parsley

Preheat oven to 180° C. Cut fish into serving-sized portions and put in a greased baking dish. Sprinkle with salt and pepper. Place lemon slices and diced onion on the fish. Mix tomato sauce, oil and dry white wine and pour over the fish. Bake for 35 minutes. Garnish with parsley. (Serves 6.)

---

# *Catfish fillet with orange rice filling*

800 g catfish fillets, washed and dried
20 m*l* salt
3 m*l* pepper
30 m*l* oil
65 m*l* orange juice

Cut through the length of the fish and sprinkle inside and outside with salt and pepper. Brush fish with oil and orange juice.

## Filling

65 m*l* oil
450 m*l* celery
40 g onion
175 m*l* water
30 m*l* lemon juice
15 m*l* grated orange peel
5 m*l* salt
250 m*l* cooked rice
125 m*l* flaked almonds

Heat oil in pan and fry onion and celery until tender. Add water, lemon juice, orange peel, salt and rice and stir until all moisture has been absorbed. Add almonds. Fill catfish with mixture and tie with string. Bake at 180° C for 35 minutes. (Serves 6.)

---

# *Smoked catfish with pasta au gratin*

### Spaghetti

95 g spaghetti
300 m*l* water
10 m*l* oil
5 m*l* salt

Break spaghetti into small pieces and boil in water with oil and salt until tender.

### Catfish

705 g smoked catfish, cut into strips of
40 mm × 15 mm × 8 mm
40 g butter
60 g chopped onion
50 g chopped mushrooms

Melt butter and fry onion, mushrooms and smoked catfish lightly until golden brown.

### White sauce

250 m*l* milk
20 g flour
20 m*l* water
2,5 m*l* salt
1 m*l* pepper
65 g cheddar cheese, grated

Boil milk. Mix flour, water, salt and pepper and add to milk. Add half the cheddar cheese, then add spaghetti and catfish. Spoon mixture into baking dish. Sprinkle cheese on top and bake at 180° C for 30 minutes until golden brown. (Serves 6.)

# *Broiled catfish with white wine*

900 g catfish fillets, washed and dried
40 m*l* oil
1 tin mushrooms (65 g)
65 g chopped onion
1 clove garlic, crushed
125 m*l* white wine
30 m*l* chopped parsley
10 m*l* salt
2,5 m*l* thyme
5 m*l* pepper

Place each fillet on a 15 cm × 15 cm piece of foil. Bring edges of foil together to form a bowl. Heat oil and fry mushrooms, onion and garlic until tender. Add white wine, lemon juice, parsley, salt, pepper and thyme and heat through. Pour the sauce over each fillet. Seal thoroughly. Bake at 180° C for 35 minutes or until done. Garnish with a sprig of parsley. (Serves 6.)

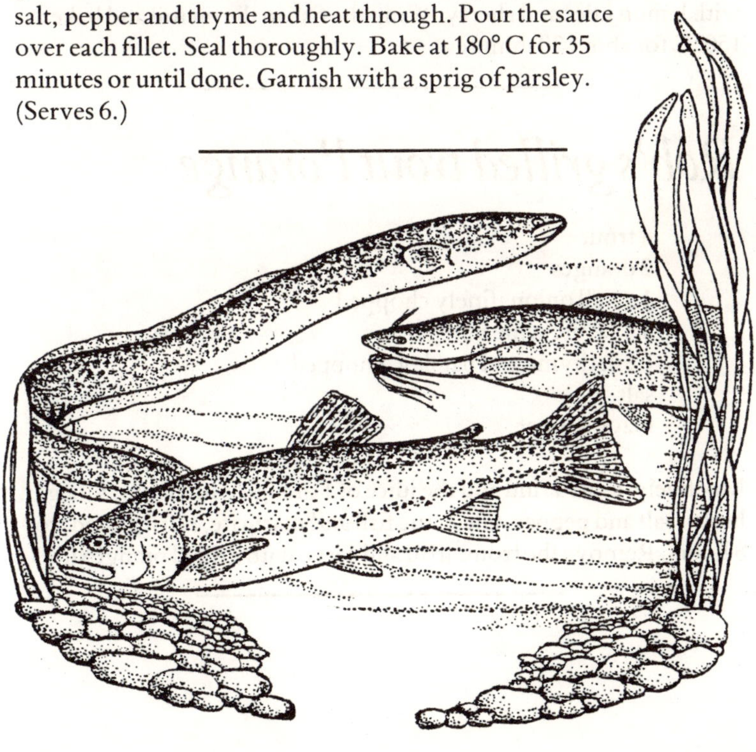

# TROUT

## *Rob's trout braai*

1 trout per person
10–15 g butter per trout
garlic salt
mixed herbs
black pepper
10–12 m*l* full cream sherry per trout
1–2 m*l* lemon juice per trout

Clean the fish well and put on tin foil. Put butter in body cavity and season skin with garlic salt, mixed herbs and black pepper. Sprinkle with lemon juice and sherry. Wrap the fish in the tin foil and bake at 150° C for about 20 minutes.

---

## *Rob's grilled trout l'orange*

4 trout
2 oranges
1 small onion, finely chopped
4 T white wine vinegar
fresh thyme and parsley, chopped
salt
black pepper

Mix finely-grated rind and the juice of 1 orange with onion, vinegar, herbs, salt and pepper. Clean the fish and marinate, refrigerated, for 8 hours. Remove the trout, grill and serve with grilled orange slices.

---

## *Baked trout with almonds*

4 trout, rinsed and dried
¼ cup melted butter
juice of a large lemon
paprika
½ cup cream
¼ cup dry white wine
½ cup flaked almonds
1 t dried dill
salt and black pepper
parsley, finely chopped

Season fish inside and out. Put in a greased, shallow baking dish, pour over the melted butter and lemon juice and dust with paprika. Grill for 5 minutes without turning until light-brown. Remove from the oven and turn the temperature to 200° C. Then pour the cream and wine over the fish and sprinkle with nuts and dill. Bake for 10–15 minutes until cooked through and sprinkle with parsley.

---

## *Pan-fried trout*

4 trout, dampened
bread *or* seasoned cornflake crumbs
salt and pepper
250 m*l* oil
250 m*l* butter

Coat the dampened trout with crumbs. Heat the butter and oil in a heavy-bottomed pan until pale brown. Reduce to medium heat. Fry trout about 4 minutes on each side at medium heat, turning once only. Drain and keep warm. Serve with new potatoes, broccoli and a salad.

An alternative to this would be to coat the trout with crumbs, wrap them in bacon strips and fry carefully for about 5 minutes on each side. (Serves 4.)

---

# *Trout with mushroom sauce*

4 boned trout
flour, tossed with salt and pepper
75 m*l* marjoram
75 m*l* oil
250 g button mushrooms, thinly sliced *or* 1 tin button mushrooms (285 g)
1 clove garlic, chopped
375 m*l* white wine, sherry *or* Pernod
125 m*l* thick cream

Dust trout with flour. Heat butter and oil and fry trout over medium heat for about 5 minutes on each side. Remove and keep warm. Put mushrooms and garlic in a pan and sauté. Add wine, turn up heat and cook for 1 minute. Add cream, remove from heat, reduce heat and return, stirring until blended. Season with freshly ground black pepper. Serve with potatoes and a salad.

# *Hannetjie's smoked trout pâté*

4 smoked trout, boned and flaked
1 onion, finely chopped *or* grated
25 m*l* melted butter
10 m*l* finely chopped parsley
5 m*l* chopped chives
50 m*l* cream

Liquidise all ingredients except the cream and butter. Add cream and butter gradually. Put in individual ramekin dishes and garnish with chopped parsley. Serve with wholewheat or crisp bread. (Serves 4.)

# White Fish

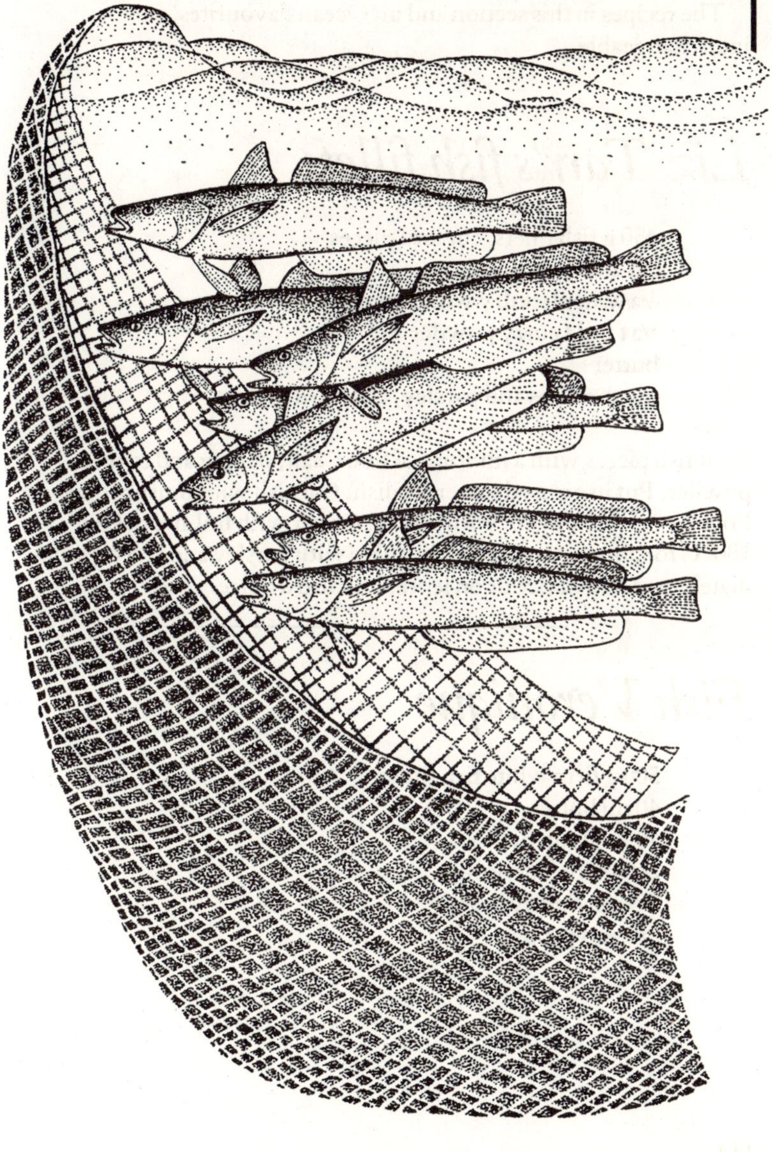

Although only a few hundred species of fish are used regularly by man, it is estimated that there are about 2 600 species of fish on our planet. White fish can be a real culinary experience. This section and 'Ocean Favourites' should encourage you to experiment with new species when available, or use frozen fish out of season.

The recipes in this section and in 'Ocean Favourites' are interchangeable.

## *Liz Tarr's fish fillets*

450 g fish fillets, cut into pieces
2 T flour
½ wineglass dry white wine
½ t mild curry powder
butter
mushrooms and parsley, chopped

Coat fish pieces with a mixture of flour, salt, pepper and curry powder. Put in greased casserole dish. Garnish with mushrooms. Pour over wine and dot with butter. Cover with foil and bake at 180° C for about 20 minutes. Sprinkle with parsley and serve immediately on a bed of rice or on lentils and brown rice.

---

## *Fish Véronique*

500 g firm white fish fillets
40 m*l* butter
40 m*l* flour
250 m*l* breadcrumbs
2 eggs, lightly beaten
salt and pepper

### Sauce

10 m*l* butter
20 m*l* flour
1 chicken stock cube
250 m*l* milk
250 m*l* sour cream
250 g seeded white grapes
pinch of paprika

Toss fish in seasoned flour, coat with egg and breadcrumbs and fry in butter over medium heat. Set aside. To make sauce: in a separate pan, heat butter and stir in flour. Gradually add milk and crumbled stock cube, stirring constantly until smooth. Add sour cream, paprika and grapes and simmer over gentle heat for 5 minutes. Place fish fillets in serving dish, pour over sauce and heat gently in oven. Garnish with parsley and lemon wedges.

---

## *Georjane Oberholzer's Cape fish pie*

250 m*l* flaked raw *or* pickled fish
6 slices bread, buttered and quartered
375 m*l* grated cheddar cheese
2 eggs, lightly beaten
250 m*l* milk
30 m*l* sherry
2,5 m*l* Worcestershire sauce
salt and pepper to taste

Grease a baking dish with butter. Put a layer of bread, fish and cheese in the dish ending with a layer of bread. Mix all the other ingredients and pour over the mixture in dish. Bake at 125° C for 1 hour. (Serves 3–4).

---

# *Jean Pote's fish bobotie*

1 kg hake fillets, cooked and flaked
1 thick slice white bread
1 cup milk
2 onions, sliced
1 T butter *or* oil
2 T curry powder
2 t sugar
salt and pepper to taste
2 T vinegar
2 eggs
6 lemon leaves *or* 3 bay leaves

Soak bread in milk and squeeze dry. Reserve remaining milk. Sauté onions in butter. Add curry powder, sugar, salt and vinegar and boil. Add fish and soaked bread. Stir in 1 beaten egg. Put in greased baking dish. Beat remaining egg and milk together and pour over fish. Stick lemon or bay leaves into mixture and bake at 190° C until egg and milk have set into a custard. Serve with boiled rice and chutney. (Serves 6.)

---

# *Fish pudding à la Jubb*

1 fresh white fish *or* haddock, skinned, boned and flaked
thick slice bread, crumbed
25 m*l* margarine
1 *or* 2 eggs
75 m*l* milk *or* fish stock
salt and pepper
12,5 m*l* chopped parsley

Steam fish, drain and retain stock. Make stock up to 75 m*l* with milk. Add margarine, heat and pour onto breadcrumbs. Mix all ingredients. Put in greased pudding bowl, cover with greased paper and steam for 3/4 hour. Serve with an egg sauce.

---

# *Sweet and sour fish bites*

500 g kabeljou *or* kingklip *or* hake *or* monk fillets, cut into strips
2 eggs
30 m*l* milk
500 m*l* fresh white breadcrumbs
60 m*l* seasoned flour
oil for deep-frying

## Sauce

1 small diced onion
1 tin pineapple pieces (410 g)
15 m*l* tomato purée
15 m*l* cornflour *or* custard
1 m*l* ground ginger
1 green pepper, sliced
15 m*l* brown sugar
125 m*l* vinegar

Coat fish in seasoned flour. Beat the eggs and milk together. Dip the fish into the egg mixture and then coat with breadcrumbs. Put on a lightly floured board and roll under palm of hand to round and lengthen the strips. Refrigerate while making the sauce. Drain the pineapple and reserve 150 m*l* of the juice. Fry the onion and pepper rings in a little oil until soft, then drain. Put reserved juice, vinegar, brown sugar, ginger and tomato purée in a pan. Blend the cornflour to a smooth paste with a little water and stir into the pan. Bring to the boil stirring constantly and simmer for 3 minutes. Remove from heat and add drained pineapple, onion and green pepper. Keep warm. Deep-fry the fish bites in hot oil for 3 minutes until golden. Arrange on serving dish and pour over the sauce. Serve with boiled rice and a green salad.

# Fish balls

1 kg filleted stockfish, cooked, drained and flaked
2 T chopped parsley
1 egg, beaten
1 onion, finely chopped
salt and pepper
2 cups cornflakes, crushed

## White sauce

1 cup milk
2 T flour
butter

Combine the fish, parsley, onion and seasoning. Make 1 cup white sauce and add to fish mixture. Refrigerate for 1 hour. Shape into balls. Roll in cornflakes, dip into beaten egg and again roll in cornflakes. Fry in oil until brown. Serve with sauce tartare or tomato sauce.

---

# Fish fillets in mushroom cheese sauce

1 kg fillets of any fried *or* steamed white fish
paprika
1 cup grated cheese
salt and pepper
1 tin cream of mushroom soup (410 g)
¼ cup sherry

Arrange fillets in baking dish and season. Mix remaining ingredients except paprika and pour over fish. Sprinkle with paprika and bake for about 15–20 minutes at 180° C.

---

## *Easy fish casserole*

1 kg white fish
1 tin cream of mushroom soup (410 g) *or* soup made from
  2 packets of mushroom soup
½ cup milk
1 tin peas (410 g), drained

Steam fish and allow to cool. Add milk and peas to soup. Pour over fish and bake at 180° C for about 25 minutes.

---

## *Fish and pineapple titbits*

1 pkt frozen fish fingers (400 g)
1 small tin pineapple pieces (410 g)
buttered toast

Cube semi-thawed fish fingers and fry in hot oil until golden brown. Coat pineapple pieces with crumbs and fry until golden brown. Mix with pieces of fried fish and pile onto hot buttered toast. (Serves 6.)

---

## *Fish and vegetable pie*

2–3 cups cooked, flaked fish
1 cup cooked mixed vegetables
600 m*l* thick white sauce
grated cheese
breadcrumbs
seasoning

Combine fish, vegetables and white sauce. Put in greased baking dish. Combine breadcrumbs and cheese and sprinkle over fish. Dot with butter and season. Bake at 190° C for about 25 minutes. (Serves 8.)

---

# *Steamed fish with raisins*

1 kg filleted fish, steamed, skinned and boned
1 lemon
1 cup hot fish stock, salted
1 cup chilled seedless raisins
¼ cup sugar
2 egg yolks, beaten
1 t chopped parsley

Arrange fish on platter and keep warm. Combine sugar, grated lemon rind and lemon juice with yolks and add fish stock gradually. Cook until thickened, stirring constantly. Add salt and parsley. Pour over fish, sprinkle with raisins and serve immediately. (Serves 6–8).

---

# *Baked paprika fish*

700 g firm white fish
2 T butter *or* margarine
¼ t paprika
1 onion sliced
juice of ½ lemon *or* ¼ cup white wine
⅛ cup water
salt, pepper and spices to taste
¼ t paprika
2–3 T sherry *or* dry red wine

Melt 1 T butter or margarine in a saucepan. Add paprika and onion and sauté. Clean fish and wash in lemon juice and water or white wine and water. Leave for 5 minutes. Cut into 60 mm pieces, put in a greased baking dish and dot with 1 T butter or margarine. Sprinkle with salt, pepper, spices, paprika and sherry. Add sautéed onions. Bake at 190° C for 30 minutes. Grill 1–2 minutes, sprinkle with paprika and parsley and serve. (Serves 4.)

---

# *Ethnic Specialities*

The J.L.B. Smith Institute and ICHTHOS are melting pots for various nationalities and cultures. These cross-cultural influences and numerous visits overseas have led to the discovery of unusual ethnic methods of preparing fish. Some are traditional staple diets while others are served during feasts or celebrations. Also included here are the traditional Malay-Dutch recipes like bobotie, bredie, curried and pickled fish.

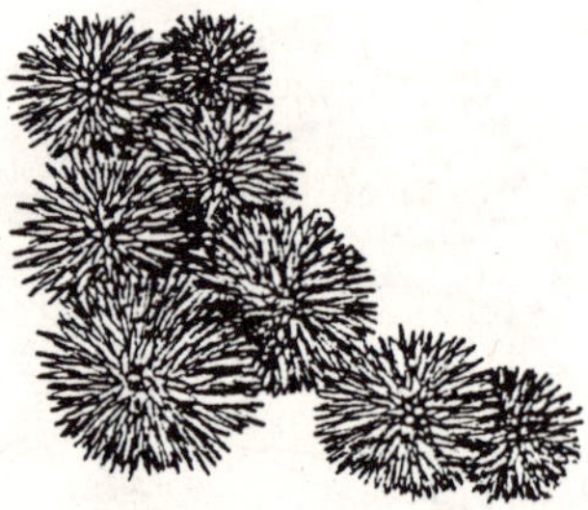

# *Liz Tarr's Malawi banana bonanza*

any white fish fillets
green bananas, sliced
onion, chopped and lightly fried
tomatoes, sliced
2 t curry powder
salt to taste

In a casserole dish layer fillets of fish, banana, onion and tomato, ending with a layer of banana. Mix curry powder and salt to taste with enough water to almost cover layers. Bake, covered, at 180° C for 30 minutes or until done.

# *Margaret Coulter's Samoan kokoda*

*Kokoda, pronounced 'kakonda', is a well-known dish on the islands of the South Pacific. It is based on the traditional staple foods of the islanders: coconuts, fruit and fish. Westerners usually serve it as an entrée with dinner, often in a goblet-type fruit dish. Be forewarned: it can become addictive!*

750 g firm, white-fleshed fish such as mullet *or* kingfish *or* bream
lemon *or* lime juice
thick coconut cream
spring onion, carrot and green pepper, finely chopped

Cut fish fillets into 1 cm cubes, sprinkle with salt and pour over 1 cup lemon juice to cover. Refrigerate until white (usually 6–12 hours, depending on kind of fish and acidity of lemons). Strain the juice and squeeze out moisture by pressing in a colander. Pour 1 cup coconut cream over fish and add carrot, spring onion and green pepper. Garnish and serve chilled but not too cold. (Serves 6.)

*Variation: Half the fish may be replaced by shelled and headed prawns.*

---

# *Tom Hecht's Argentine ceviche*

1 firm white fish, boned and cut into 18 mm cubes
salt and pepper
3 onions, thinly sliced
2 T fresh lemon juice
red peppers
bay leaf

Season fish with salt and pepper, mix with onion, put in a glass dish and pour over fresh lemon juice to cover. Garnish with red peppers and bay leaf and refrigerate for a minimum of 24 hours. Serve with wholewheat bread and cold beer or chilled white wine.

---

# *Ofer Gon's Hawaiian fish teriyaki*

*Most marine fish can be used for this dish.*

Teriyaki sauce for 6 fish *or* portions
6 T thick soy sauce
6 T dry *or* medium sherry
1 T brown sugar
6 T sake (rice wine) *or* dry red *or* white wine
1 clove garlic, finely chopped
about 1 T grated fresh ginger root

Mix the ingredients in a pot, bring to the boil and pour over fish. Marinate fish in the sauce for about 15–20 minutes, turning them over every few minutes. Grill 5–10 minutes on each side, depending on size of fish, and baste frequently with sauce. If whole fish is used, make 2–3 transverse cuts on body.

# *Taramasalata*

*This Greek fish roe pâté is served as an hors d'oeuvre.*

250 g smoked fish roe
125 g sliced white bread, moistened with a little water
250 m*l* olive oil
juice of a large lemon
2 cloves garlic, minced
1 small onion, finely chopped
freshly ground black pepper
black olives

Skin roe and break into lumps. Soak in water for about an hour and drain. Squeeze bread dry and put roe and bread in a blender. Switch on blender and add oil, very slowly, in a thin stream, as if preparing mayonnaise. When mixture is a pale pink purée, add lemon juice to taste and more oil if preferred. Add garlic. Stir in chopped onion and freshly ground black pepper to taste. Spoon mixture into serving dish and garnish with black olives. Serve with hot buttered toast and black olives. (Serves 8.)

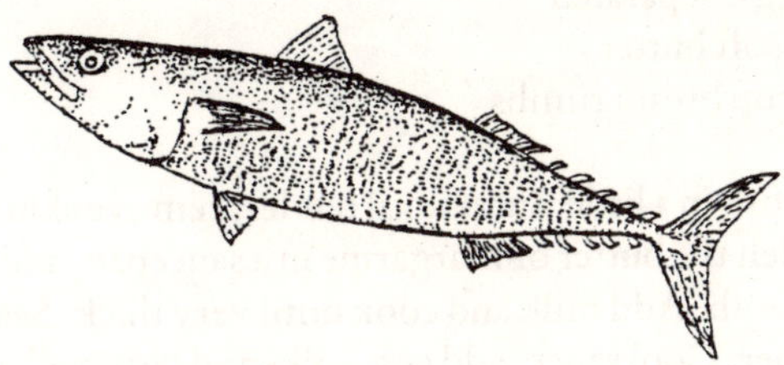

# *Chambo nkopola*

*This dish comes from the southern shores of Lake Malawi.*

4 fillets of bream
1 T flour
salt and pepper
1 T butter
2 naartjies
1 T vinegar *or* white wine
1 T chopped parsley

Coat fillets of fish with seasoned flour. Heat butter and fry fillets gently for 4–5 minutes on each side without browning the butter. Remove and keep warm on serving dish. Grate a little rind from 1 naartjie and add to frying pan with juice of 1 naartjie, vinegar and chopped parsley. Bring to the boil and pour over fillets of fish. Garnish with segments of naartjie and parsley.

# *Margaret Jackson's fiskibudingur*

*This fish soufflé comes from Iceland.*

500 g kob *or* stockfish *or* red line fish
½ cup butter *or* margarine
1 cup sifted flour
1¼ cups milk
1 t salt
few grains of pepper
6 eggs, separated
1 T soft butter
½ cup breadcrumbs

Simmer the fish in a little water until tender, remove skin and bones and flake. Melt the butter or margarine in a saucepan, add flour and stir until smooth. Add milk and cook until very thick. Season with salt and pepper. Cool sauce, add egg yolks and beat well. Combine the fish and sauce and fold the beaten egg whites into the fish mixture. Pour mixture into a buttered 1,5 *l* casserole dish coated with crumbs. Sprinkle bread crumbs over the top and bake at 190° C for 55 minutes. (Serves 6–8.)

---

# *Margaret Jackson's bachalhau do Céu*

*This heavenly codfish dish comes from Portugal.*

1 kg fresh *or* frozen codfish *or* other large fish such as kob, cut into serving portions
1 kg potatoes, uniform size, boiled in jackets
2 T butter
¼ cup milk
2 onions, sliced
¼ cup olive oil

8 hard-boiled eggs, sliced
béchamel sauce (see preceding recipe)
1 egg yolk, lightly beaten

Drain, cool and peel boiled potatoes. Slice about ¾ of the potatoes, mash remaining ¼ and season with butter and milk. Fry onions in hot oil in a 300 mm saucepan until transparent. Add fish and cook over low heat for 30 minutes. Grease a 1,5 *l* casserole. Put a layer of fish in the casserole, then a layer each of sliced potato, egg and béchamel sauce. Continue layers in order given using all ingredients. Top with mashed potato. Brush with egg yolk. Bake at 180° C for 45 minutes or until potato is golden brown. (Serves 6–8.)

## *Margaret Jackson's Nigerian ewa and dodo*

1 tin tuna (210 g) *or* 2 small tins shrimps (200 g each)
2 cups black-eyed peas, dried
4 cups boiling water
½ cup onion, chopped
1 large tomato, chopped
1½–3 t dried red pepper, crushed
2 T tomato paste *or* sauce
3 T oil
3 large bananas, cut into 10 mm slices
½ t salt
1 cup oil for frying

Wash peas in cold running water. Cover with boiling water and boil for 2 minutes. Remove from heat and soak for 1–2 hours. Simmer in the same water until tender. Add onion, tomato and red pepper to peas and cook for 15 minutes. Add tomato paste, oil and fish, cover and simmer for 10 minutes without stirring. Remove cover, stir and simmer for another 5 minutes. Sprinkle bananas with salt. Deep-fry in oil until golden brown. Drain on absorbent paper. Serve black-eyed pea dish (*ewa*) with bananas (*dodo*). (Serves 6.)

# *Margaret Jackson's baked fish à la Moscow*

6 fillets of fish, each weighing 180–250 g
½ t salt
½ t pepper
½ cup flour
2 T tomato paste *or* 1 egg yolk
½ cup oil
2 cups béchamel sauce
½ cup chopped onions, sautéed
½ cup sliced mushrooms, sautéed
1 cup cooked crab *or* lobster meat, flaked
3 hard-boiled eggs, quartered
3 medium-sized potatoes, sliced and fried
¼ cup grated cheese

## Béchamel sauce

¼ cup butter
¼ cup flour
½ t salt
⅛ t pepper
1½ cups hot chicken stock
1 cup scalded milk
2 T tomato paste *or* 1 egg yolk, lightly beaten

Melt butter and add flour, salt and pepper. Gradually add chicken stock and milk. Cook over low heat for 3–5 minutes. Add tomato paste to give flavour and colour or add egg yolk for yellow sauce.

Coat each fillet with a mixture of salt, pepper and flour and fry in hot oil until brown on both sides. Put fillets in a baking dish of about 250 × 350 mm over which a thin layer of béchamel sauce has been spread. Add the onions, mushrooms, crab or lobster meat and eggs. Cover with fried potato slices. Pour remaining sauce on top of potatoes and sprinkle with cheese. Bake at 180° C for 15–20 minutes or until heated through. (Serves 6.)

# *Margaret Jackson's Italian fish à la spetsiota*

1 kg red snapper fillets
2 t salt
¼ t pepper
2 T lemon juice
2 cups tinned tomatoes, drained
½ cup white wine
⅓ cup olive oil
½ cup minced parsley
2 cloves garlic, crushed
⅓ cup dry breadcrumbs
½ cup tomato juice

Cut ready-to-cook red snapper into 12 pieces and rub with salt and pepper. Combine all other ingredients except breadcrumbs and tomato juice. Pour into a casserole or baking dish. Place fish in sauce. Bake uncovered at 180° C for 1½ hours. (Serves 6.)

---

# *Carolynn's Chinese fish balls in sweet and sour sauce*

500 g cooked 'dassie' (blacktail), boned and flaked
salt and pepper
1 egg
120 g cornflour
oil for deep-frying
4–5 spring onions, chopped

### Sauce

½ green and ½ red pepper, seeded and chopped
½ medium-sized onion, chopped
oil
pinch ground ginger
2 T vinegar
4 T sugar
2 T tomato sauce *or* tomato paste
2 T cornflour
5 T soy sauce
1 T brandy
25 m*l* water

Season fish with salt and pepper. Add egg and cornflour, then stir until the mixture is smooth. Shape into small balls. Make the sauce by frying the peppers and onion in very little oil, sprinkled with ginger. Mix the vinegar, sugar, tomato sauce, cornflour, soy sauce and brandy and blend with the water. Add this mixture to the fried vegetables. Simmer gently for 5 minutes, stirring constantly and adding more water to thin down if necessary. Keep warm. Deep-fry the fish balls in oil until golden. Drain on absorbent paper and transfer to a hot serving dish. Cover with sweet and sour sauce. Garnish with spring onion. (Serves 4.)

---

# *Traditional Jewish gefilte fish*

*This recipe originates from Shiknazi communities in Poland, Romania and Russia.*

1 kg trout *or* common carp *or* white fish
2 onions, grated
1 T matza meal
2 eggs
½ cup water
1 carrot, chopped
salt and pepper to taste
celery

Skin fish, set skin aside and mince or process the remaining flesh. Mix raw fish with 1 grated onion, salt, pepper, matza meal, eggs and water. Put fish skin and odd parts in a pot. Add remaining onion, carrot and celery. Shape fish mixture into balls, put in pot, and cover with water. Cook covered until water boils. Uncover and simmer for 1½ hours. Shake occasionally to prevent fish from sticking to the bottom of pot. (Serves 6.)

---

# *Jewish fish blintzes*

250 g mackerel, flaked
1 cup cottage cheese
1 egg
2 T sugar
¾ t salt
¼ t cinnamon
1 cup sour cream
½ cup cherry preserves

## Blintz batter

¾ cup flour
2 beaten eggs
¼ t salt
oil
1 cup milk

To make blintzes, mix flour and ¼ t salt. Beat milk and eggs. Gradually add egg and milk mixture to flour and stir until smooth. Heat oil in saucepan. Pour 2 T batter into pan and tip to let batter cover bottom. Fry 3–4 minutes or until bottom of 'pancake' or blintz is brown. Remove and continue process until about 12 blintzes are done. Add cottage cheese, egg, sugar, salt and cinnamon to fish. Place 2 T of fish mixture on browned side of each blintz. Spread to within 3 cm of edges. Roll blintz, folding in sides. Fry in greased saucepan for 5 minutes. Turn and fry for another 5 minutes. Drain. Top with sour cream and preserves. (Serves 6.)

---

# Portuguese party fish

*There are two methods of preparing this dish: by baking the fish in a white sauce (to which other ingredients have been added) or by cooking dry with the added ingredients sprinkled over the fish in the baking dish. Both methods are explained below. Either cutlets or fillets may be used.*

5 kg fish cutlets *or* fillets
250 g minced gherkins *or* pickled cucumber
3,75 *l* white sauce
6 bay leaves
3 kg onions, finely chopped
4 kg chopped tomatoes
250 g sugar
1,5 kg margarine, if using dry method *or*
500 g margarine, if using sauce

## Method 1: Using sauce

Sauté the onions in 500 g margarine and add to the white sauce. Add the rest of the ingredients and simmer for 20 minutes. Put fish in lightly greased baking dishes and pour the sauce over so as to almost cover the fish. Bake at 220° C for 45 minutes.

## Method 2: Dry method

Put fish in greased baking dishes. Mix onion, tomato, gherkin and sugar and sprinkle over fish. Add cloves and bay leaves. Dot with margarine. Bake at 220° C for 25 minutes. Serve with one thin slice of raw tomato on each portion. (Serves 20.)

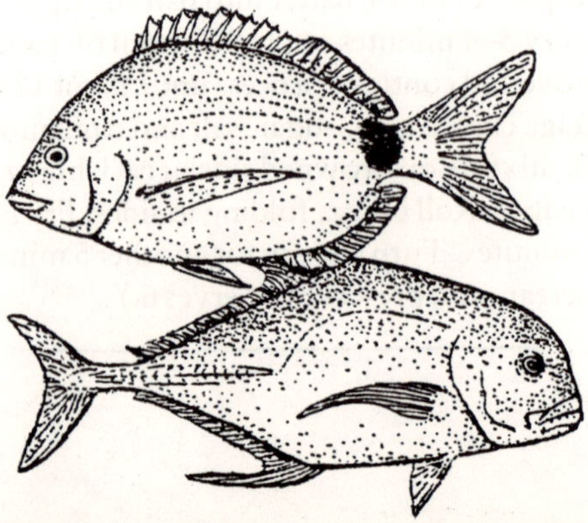

# *Korean fish and steak dish*

500 g fish fillets, cut into bite-sized pieces
225 g steak, cut into thin strips
3 celery stalks, chopped
1 small turnip, chopped
2 carrots, sliced
2 green chillies, finely chopped
5 spring onions, chopped
3 T soy sauce
125 g mushrooms, sliced

## Marinade

2 t sugar
2 T soy sauce
2 T oil
1 T ground sesame seed
1 clove garlic, crushed

Combine all marinade ingredients and beat thoroughly. Arrange beef strips in a shallow bowl, baste well with marinade and set aside for at least 20 minutes, turning occasionally. Preheat oven to 180° C. Arrange beef strips at the bottom of a medium-sized casserole dish. Cover with a layer of mushrooms, then celery, turnip and carrots.

Arrange fish fillets on top and sprinkle with spring onion and chilli, reserving some for garnishing. Pour over just enough water to reach halfway up the mixture, then add soy sauce. Bring to the boil on top of stove, cover and place in oven. Cook for 15–20 minutes or until fish flakes easily. Garnish with remaining spring onion and chilli. (Serves 4–6.)

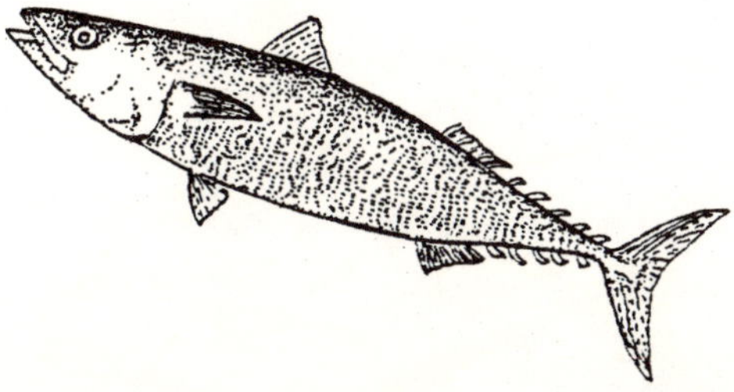

# *Liz's Fijian-style sea barbel*

800 g fillets of sea barbel *or* firm white-fleshed fish
175 g desiccated coconut
475 m*l* milk
250 m*l* cream
juice of 1 lemon
salt and pepper

Combine milk, cream and coconut, bring to the boil and simmer for 10 minutes. Allow to cool. Sprinkle fish fillets with lemon juice, salt and pepper. Put fillets in greased baking dish, pour over coconut mixture and bake at 190° C for 30 minutes. (Serves 4–6.)

# An Occasional Fancy

This section accommodates species of fish that appear only once in the book; from the national fish species (galjoen) to the more exotic carp which is much favoured by the European communities. Red snapper is found in the estuaries of the Transkei and provides the sport fisherman with exciting challenges as does the freshwater tigerfish. Another imported angling species is the bass, and conservationists are encouraged to fish for this tasty predator. Tilapia or bream is one of the most palatable of the freshwater fish. They are cultured world-wide in artificial fish ponds to meet the ever-increasing demand.

# *Glenn's Okavango bream*

Catch one large tilapia (or bream). Scale and fillet it over a slow burning fire during a beautiful Okavango sunset. Lightly season with Fondor, salt and pepper on each side. Place the fish over the coals in a regular braai grid. Allow to cook for about 5 minutes on each side. The fish meat will then flake off in bite-sized chunks. Now you can enjoy one of the best tasting dinners the Okavango has to offer.

---

# *John's Port St Johns sea barbel*

6 sea barbel fillets, cut into 4 cm pieces

## Coating

2 cups mealie meal
½ cup flour
2 t salt
1 t pepper
2 t thyme
2 t parsley
1 t coriander

Mix coating ingredients and put in a plastic bag. Drop pieces of fish separately into the bag and shake vigorously until the pieces are well coated. Heat plenty of oil in a pan and fry the pieces until crispy golden, turning only once. Serve with a green salad, crusty bread and cold beer. The bag of coating may be kept and used again.

---

# *Galjoen à la K.Z.*

Try to catch a thick, fat galjoen of over 2 kg (otherwise disregard the rest of this recipe!). Bleed, gut and scale. Slice through the skin of the fish on both sides (to fillet). This will make it easier to serve when it is cooked. Salt and pepper the fish on both sides and put it in a baking dish. Add 1½ cups of dry white wine, the juice of 1 lemon and sprinkle with parsley, mint and sweet basil. Fry three large onions in oil and when brown place in the baking dish with the galjoen. Bake at 180° C for 40 minutes. Serve with a white sauce and baked potatoes.

---

# *Kabeljou thermidor*

750 g kabeljou, skinned, boned and cut into serving portions
125 m*l* milk
1 tin cream of mushroom soup (410 g)
25 m*l* dry sherry
25 m*l* brandy
250 m*l* grated cheddar cheese
salt and pepper

Season fish pieces, put in baking dish, pour over milk and bake uncovered at 175° C for 20 minutes. Strain the liquid. Combine undiluted soup, sherry, brandy and baking liquid in a saucepan. Bring to the boil, stirring constantly. Pour over the fish, sprinkle with grated cheese and bake for 5 minutes or until cheese has melted and is turning golden. (Serves 4–6.)

---

# *Jean Pote's bass in cheese and wine sauce*

4 fillets of bass
seasonings to taste
30 g butter *or* margarine
125 m*l* dry white wine
2–3 T grated cheese
300 m*l* cheese sauce
croutons

Season fillets and put in a greased baking dish. Top with butter and wine, cover with foil or lid and cook at 190° C for 30–35 minutes. Meanwhile, make the cheese sauce, using slightly less milk than in recipe (i.e. less the amount of wine used). Drain fish, put on a hot serving dish and keep warm. Combine fish juices with cheese sauce. Coat fish with sauce. Top with cheese and put under hot grill for 1–2 minutes. Garnish with croutons.

## Cheese sauce

30 g butter
30 g flour
300 m*l* milk
seasoning, including mustard
90–180 g grated cheese

Melt butter in a pan, stir in flour and cook for a few minutes. Remove from heat and gradually add milk. Bring to the boil and cook until thickened. Add seasoning and cheese, then heat gently, stirring until smooth.

---

# *Jean Pote's bream sauvière*

4 bream (or tilapia) *or* bass fillets
500 g mushrooms
180 g margarine
seasonings to taste
dash of cayenne pepper
juice of ½ lemon
4 capers
1 onion, finely chopped
4 T white wine
sprinkling of flour

Sprinkle lemon juice over bream fillets and marinate for half an hour. Remove fillets from juice and dry carefully. Season to taste, sprinkle with flour and fry in margarine until golden. Sauté onions and mushrooms in a little margarine. Add capers. Sprinkle with flour and add the wine. Stir until slightly thickened. Put bream on hot serving dish and pour mushroom mixture over. Serve with lemon slices and boiled potatoes.

---

## *Kevin Cochrane's tilapia and asparagus*

500 g tilapia (or bream) fillets
1 tin (410 g) asparagus cuts, drained (retain liquid)
1 T oil
½ t mixed herbs
salt and pepper to taste

### White sauce

1 T margarine
1 T flour
liquid from asparagus made up to 250 m*l* with milk
salt and pepper to taste
4 T grated cheese

Heat oil in a pan and fry fillets for about 5 minutes on each side. Drain and put in an ovenproof dish. Cover fillets with the asparagus cuts. Sprinkle with salt, pepper and herbs. Make a white sauce with margarine, flour and liquid. Season to taste and pour sauce over fish and asparagus. Sprinkle with cheese and bake at 200° C for about 20 minutes or until cheese is melted and browned.

---

## *John Minshull's tiger rounds à l'Afriki*

1,5 kg fresh tigerfish
2 cooking pots
2 tomatoes
2 onions
herbs to taste
cooking oil
6 small sticks

Decapitate, scale, fin and gut the fish. Then cut the body transversely into rounds of 4 cm thick. Put rounds in a pot, cover with water and bring to the boil. After boiling for 10–15 minutes most of the fish oil should have floated to the top. Discard the fish oil and water. Dry the rounds on a towel. Place a little cooking oil in another dry pot and carefully place the sticks across the bottom. Place the dried rounds on top of the sticks. Add chopped tomato and onion with a sprinkling of herbs and cook until the vegetables are well done. The flesh of the fish will fall away from the bones in firm flakes or blocks. (Remember not to eat the sticks – they prevent the rounds from sticking to the bottom of the pot!)

---

# *Alan Whitfield's grunter delight*

spotted grunter with a total fillet weight of 800–900 g
1 heaped T butter
4 carrots, finely diced
5–7 mushrooms, sliced
2–3 onions, finely sliced
250 g frozen peas
salt, pepper and mixed herbs to taste
1 tin tomato paste (110 g)
6 T cream
1 cup white wine

Fillet spotted grunter into portions less than 5 cm thick. Put in a large greased casserole dish and season with salt, pepper and mixed herbs. Add butter, carrots, mushrooms, onions and peas. Mix tomato paste and cream, add wine, mix and pour over fish. Bake at 180° C for about 30 minutes. Serve with fresh bread rolls.

---

## *Spiced carp*

2 medium-sized carp, cleaned and cut into segments, including heads
½ cup white sugar
1 large carrot, peeled and diced
2 medium-sized onions, peeled and quartered
½ t salt
1 t ground cloves
1 t cinnamon
½ t paprika
½ t pepper

Put ingredients in a medium-sized saucepan. Cover with 1 *l* water and bring to the boil. Simmer for about 1 hour. Taste fish sauce. If too sweet, add a few drops of lemon juice. Remove onions and allow to cool. Transfer contents to a large bowl and refrigerate until fish sauce has jelled. Garnish with parsley and serve. (Serves 4–6.)

---

## *Periwinkles in spicy tomato sauce*

12–18 periwinkles per person
2 onions, peeled and sliced
2 cloves garlic, crushed
oil for frying
8 tomatoes, peeled and sliced
2 potatoes, peeled and sliced
salt and pepper
pinch allspice
2 bay leaves
1 T fruit chutney
1 T lemon juice
1 t brown sugar

Boil the periwinkles for 8–10 minutes, remove from shells and clean. Fry the onion and garlic in oil until just beginning to brown, then add tomatoes, potatoes, seasoning, allspice and bay leaves. Simmer for 15 minutes, then add chutney, lemon juice, sugar and periwinkles. Cook for another 5 minutes over a gentle heat or until the potato slices are done. Serve immediately.

---

## *Old Cape Dutch fish rissoles*

1 kg dageraad, skinned and filleted
1 onion, finely chopped
½ T butter
1 T oil
2 slices bread, crusts removed
milk
1 t parsley, finely chopped
2 eggs, beaten
2 cream cracker biscuits, finely crushed
½ t nutmeg
½ t cayenne pepper
½ t salt
1 cup breadcrumbs
oil for frying

Fry onion slowly in butter and oil. Soak bread in milk or water and squeeze dry. Mince fish and mix with parsley, one beaten egg, bread, onion, biscuits and seasoning. Shape mixture into balls. Roll each ball in beaten egg and then in breadcrumbs. Fry the balls in hot oil until crisp and golden brown, then drain on absorbent paper. Transfer to a hot serving dish and serve immediately. (Serves 4–6.)

---

# *Grunter fish bites*

350 g cooked grunter, flaked
350 g creamed mashed potatoes
2 T chopped parsley
¾ t anchovy essence
1 heapted t chopped capers
1 egg, beaten
30 g butter, melted
1 t lemon juice
½ t curry powder
1 clove garlic, crushed
salt and freshly ground black pepper to taste
2 beaten eggs, to coat
100 g dry white breadcrumbs, to coat

Combine all the ingredients except those for coating. Mix well and season to taste. Refrigerate the mixture for 1–2 hours. Lightly flour a working surface, turn fish mixture out onto it and divide into 12 piles. Shape each pile into a neat fish cake, dip in beaten egg and then in breadcrumbs, coating them completely. Shallow-fry the fish cakes in equal quantities of oil and butter until golden brown on both sides. Drain on absorbent paper and serve immediately with a sauce tartare. (Serves 4.)

# Starters, Soups, Salads and Sauces

# STARTERS

## *Seafood cocktail*

½ kg of any white fish
2 T tomato sauce
6 T mayonnaise
1 t Worcestershire sauce
1 tin peas (410 g)
freshly ground black pepper
salt to taste
paprika
parsley and lettuce for garnishing

Season and steam fish and allow to cool. Remove skin and bones and flake. Mix the tomato sauce, mayonnaise, Worcestershire sauce and pepper. Add to drained peas and fish and mix lightly. Sprinkle with paprika and serve each portion on a crisp lettuce leaf. Garnish with parsley.

# *ICHTHOS oyster soup*

1 tin oysters in liquid (104 g)
2 × 410 g tins cream of chicken soup
2 T dry sherry

Drain oysters and retain 2 T liquid. Purée oysters in liquid. Make chicken soup with slightly less water for a thicker soup. Mix all ingredients, heat and serve with a sprinkling of parsley and plenty of crusty French bread. (Serves 6.)

---

# *Vietnamese fish soup*

350 g fillet of firm white-fleshed fish, cut into 5 cm pieces
1 tin bamboo shoots (450 g), cut into 4 cm slices
½ pineapple, diced
3 T tamarind juice *or* boil ½ T tamarind in 6 T water until reduced by half and strain
2 T oil
1 t sugar
250 m*l* water
½ t salt
large pinch each of ground ginger, white pepper and chilli powder
juice of ½ lemon
½ T chopped mint

Fry bamboo shoots and pineapple in oil for a few minutes. Sprinkle with sugar and tamarind juice or water. Boil water and salt together, add fish pieces, ginger, pepper, chilli powder and lemon juice. Add bamboo shoots and pineapple and simmer over low heat for 5–6 minutes. Sprinkle with mint before serving. (Serves 4.)

---

# *Margaret Crampton's moules marinières*

75 mussels
4 large onions, chopped
garlic to taste
50 g butter
3 T flour
sufficient boiling water to almost cover mussels
1 cup wine
salt and pepper
chopped parsley

Drop mussels into boiling water. Leave until they have opened. Reduce heat and simmer for a few minutes. Remove mussels and transfer on a half shell to slow cooker or other suitable container. Reserve liquid. Cook onions in butter until transparent and tender. Add flour to make a roux. Add the liquid from the mussels and the wine until soup is of a good consistency. Add salt, pepper and chopped parsley. Add soup to the mussels and keep hot until ready to serve. Serve with thinly sliced brown bread and butter. (Serves 3 or more.)

---

# *Swedish fish soup*

340 g firm fish fillets, cut into serving portions
15 g butter
1 onion, chopped
3 medium-sized potatoes, diced
1,75 *l* fish stock
salt and freshly ground black pepper
1 bay leaf
120 m*l* thick cream
2 T chopped dill
2 T chopped parsley
2 T chopped chives

Melt butter in a large pan and fry onion until tender. Add potatoes and stock and simmer for 10 minutes. Add fish pieces, bay leaf, salt and pepper. Simmer until fish and potatoes are cooked. Remove from heat and gently stir in cream. Adjust seasoning and slowly bring to the boil. Stir in herbs and serve immediately. (Serves 4.)

---

# *Fish and onion soup*

250 g hake fillets, cut into serving portions
1 onion
oil for frying
55 m*l* chopped chives
500 m*l* fish *or* onion stock
2 m*l* thyme
1 bay leaf
125 m*l* milk
125 m*l* cream
salt, pepper and nutmeg

Fry fish and onion in a large pan. Add stock, thyme and bay leaf and simmer for 10 minutes. Remove from heat and add warmed milk and spices. Stir in cream. Simmer gently for 10 minutes and serve immediately. (Serves 4.)

---

# *Tom Pike's fish soup*

head and bones of any large freshwater or marine fish,
preferably tilapia, kob, musselcracker *or* grunter
1 large onion, chopped
1,75 *l* water
salt and pepper
½ cup fresh milk *or* a little milk powder
cornflour
shallots, chopped

Boil head and bones with onions, salt and pepper until the meat softens and separates from the bone. Strain to remove the bones, then pour the onions and meat back into the fluid. Scrape loose meat off the bones and mix into soup. Add milk or milk powder. Thicken with cornflour and sprinkle with shallots. Serve with toasted bread. (Serves 4.)

---

# *Granny's fish soup*

Head and bones of any large fish
3 carrots, sliced
2 large tomatoes, skinned and sliced
2 T curry powder
2 t lemon juice *or* vinegar
1 t sugar
1 bay leaf
salt to taste
3 boiled potatoes, sliced
1 large onion, sliced

Put head and bones in a saucepan with sufficient water. Simmer for 1 hour and strain. Brown onion in a little fat. Add curry, sugar, lemon juice, potato slices, fish stock, tomatoes, carrots and bay leaf. Simmer until vegetables are tender.

---

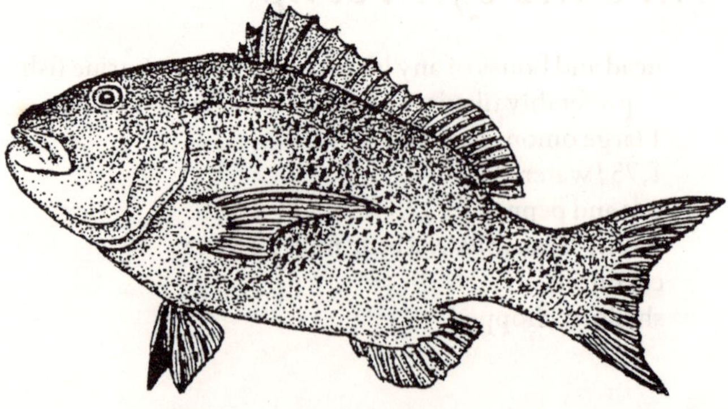

## *Seafood vinaigrette*

1 *l* fresh mussels
150 g prawns
150 g fresh, frozen *or* canned crab meat, flaked
4 new potatoes, cooked and diced
1 dessert apple, diced
1 dill cucumber, diced
1 cup broad beans *or* peas, cooked
4–5 T olive oil
2 T wine vinegar
1 clove garlic, crushed
sugar to taste
salt and freshly ground black pepper
3 cups mayonnaise

### Garnish

120 g prawns, shelled
lettuce hearts
4 hard-boiled eggs, quartered

Cook the mussels and remove from shells. Shell the prawns, then put all the seafood in a large bowl. Add the potato, apple, cucumber and beans or peas. Beat the oil and vinegar with the garlic, sugar and seasoning and work this well into the seafood mixture using two forks so as not to break it up. Adjust seasoning. Leave the mixture to blend for 20 minutes, then fold in enough mayonnaise to bind the fish and vegetables. Pile onto a serving dish, coat with more mayonnaise and garnish. (Serves 4.)

# *Mr Fick's seafood salad*

1 tin tuna (200 g)
1 tin smoked oysters (105 g)
1 tin shrimps (105 g)
1 medium-sized tomato, diced
½ cup diced cucumber
salt and freshly ground black pepper

## Dressing

1 cup mayonnaise
2 T tomato sauce
3 T fresh lemon juice
1 T fresh orange juice
2 T sugar

## Garnish

chopped chives
few black olives

Toss all the salad ingredients in a bowl and season with salt and pepper. Prepare the dressing by blending all the ingredients, pour half the dressing over salad and toss lightly. Allow to stand for 30 minutes before serving. The balance of the dressing may be served on the side. Garnish with chives and olives.

---

# *Filipino fish salad*

800 g firm white fish fillets, cut into 2 cm cubes
1 clove garlic, finely chopped
150 g desiccated coconut
6 T milk
¼ t turmeric
bunch of chopped chives
4 tomatoes, diced
1 lettuce (½ shredded, ½ in loose leaves)

### Marinade

juice of 3 lemons
1 onion, thinly sliced
3 t chopped root ginger
1 t salt
½ t black pepper

Marinate fish cubes in refrigerator overnight. Purée coconut, garlic, milk and turmeric. Stand for 1 hour. Squeeze firmly through muslin, removing all the milk. Stir coconut milk into marinated fish, then add tomato, chives and shredded lettuce. Arrange lettuce leaves on serving dish and spoon fish mixture into centre of dish. (Serves 4.)

---

## *Fish and vegetable salad*

250 g cooked fish, diced *or* leftover fried fish sticks
2 cups cooked potatoes, diced
½ T grated onion
French dressing or mayonnaise
½ cup cooked carrots, diced
½ cup cooked green peas
2 hard-boiled eggs, diced
1 gherkin, finely chopped

Combine all ingredients. Moisten with French dressing or mayonnaise and toss lightly until well coated. (Serves 4–6.)

---

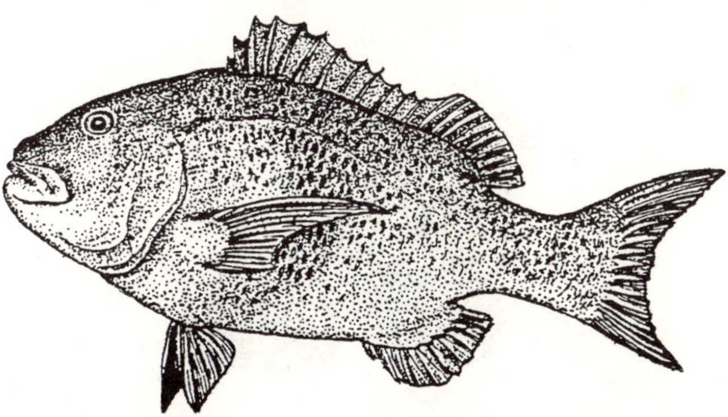

# *Linda's fish and avocado salad*

3 cups cooked fish of any kind, flaked
2 avocados, diced
2 apples, diced
lemon juice
100 m*l* mayonnaise
3 T fresh cream
lettuce leaves

Sprinkle avocado and apple with lemon juice. Combine with fish, mayonnaise and cream. Serve on lettuce leaves as a salad or starter.

---

# *Fish and macaroni salad*

2 cups cooked fish, flaked
1 cup cooked ham, diced
2 cups boiled macaroni
½ cup lemon juice
½ cup cream *or* mayonnaise
salt and pepper to taste

Combine lemon juice with cream or mayonnaise. Add remaining ingredients. Serve as a salad or starter.

---

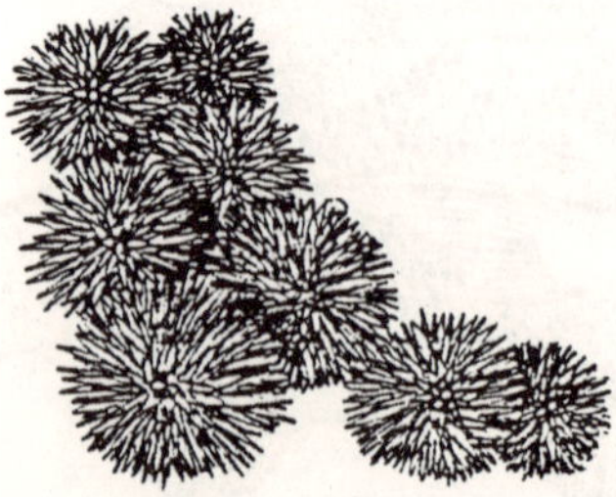

## *Carolynn's seafood sauce*

500 g cooked chilled fish *or* shellfish, cubed
50 m*l* mayonnaise
1 m*l* castor sugar
30 m*l* whipped cream
12,5 m*l* tomato sauce
12,5 m*l* chopped onion
2 m*l* Worcestershire sauce
salt, pepper and Tabasco sauce to taste

Combine mayonnaise, castor sugar, whipped cream, tomato sauce, chopped onion and Worcestershire sauce. Season with salt, pepper and Tabasco sauce. Toss with fish or shellfish.

---

## *Fish and ham spread*

equal quantites of cooked, flaked fish and finely chopped ham
mayonnaise
salt and cayenne pepper to taste

Combine fish and ham. Add sufficient mayonnaise to bind. Season with salt and cayenne pepper. Spread on crisp rolls. (1 cup serves 2.)

---

# *Cream of mushroom sauce*

1 T butter
salt and pepper
1½ cups milk
1 T flour
½ cup cooked *or* tinned mushrooms

Melt butter, blend in flour, add seasoning and slowly stir in milk. Keep stirring until sauce begins to bubble. Add mushrooms and serve.

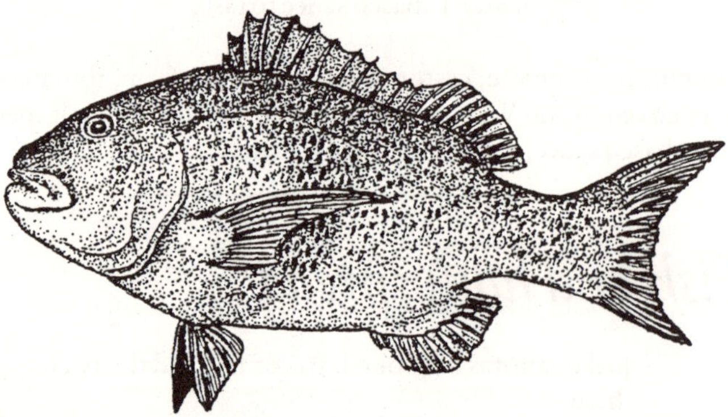

# *Sauce mexicaine*

2 chopped leeks, white parts only
salt and pepper
1 t sugar
2 large tomatoes, chopped
1 large green pepper, chopped
2 t paprika
1 T white wine

Combine all ingredients and simmer for 10 minutes.

# Onion sauce

2 onions, chopped
salt and pepper
½ t butter
¾ cup water *or* fish stock

Cook onion in water or stock until tender. Rub through sieve, add seasoning and butter and reheat. Add a little extra water or stock if too thick.

---

# Hard-boiled egg mayonnaise

yolks of 2 hard-boiled eggs
1 t sugar
2 T cream
1 T vinegar *or* lemon juice
salt and pepper
¼ t dry mustard

Pound egg yolks until fine, then add seasoning, mustard and sugar. Gradually beat in cream and then vinegar until smooth. Refrigerate.

---

# Cold parsley dressing

2 T finely chopped parsley
1 T finely chopped spring onions
½ cup salad oil
1 t salt
⅛ t pepper
2 t vinegar

Mix well and refrigerate.

---

# *Fish fingers sauce*

1 egg yolk, beaten
2 t fish paste
½ t dry mustard
1 cup mayonnaise
vinegar and pepper to taste
1 t onion juice
1 clove garlic

Combine all ingredients and mix well. Refrigerate for about 24 hours. Remove garlic before serving.

---

# *Tomato sauce*

2 large tomatoes, sliced
1 T butter
1 T chopped parsley *or* 1 t mixed herbs
salt, pepper and sugar to taste
1 T flour
2 slices onion

Simmer tomato with onion until tender. Blend in a processor or rub through a sieve. Season to taste. Melt butter and stir in flour until blended. Add tomato purée gradually, stirring constantly until sauce is smooth and has boiled. Add herbs or parsley.

---

# *Hollandaise sauce*

yolks of 2 eggs
½ cup butter
1 T lemon juice
salt and pepper to taste
½ cup boiling water

Cream butter. Add egg yolks singly, beating constantly. Add lemon juice and seasoning and beat until light. Slowly add boiling water, beating constantly.

# *Sauce tartare*

½ cup mayonnaise
1 onion, diced
2 t finely chopped pickled cucumber
1 t chopped green olives
½ t minced capers
1 t minced parsley
vinegar to taste

Add onion, cucumber, olives, capers and parsley to mayonnaise and blend well. Gradually add vinegar to taste.

# *Glazing oil*

½ cup oil
4 cm ginger root, chopped
spring onions, chopped

Simmer ginger and onion in oil for about 10 minutes. Strain and keep oil as an addition to various fried prawn and shrimp dishes.

# *Soy and spring onion dip*

4 spring onions, chopped
1 ginger root, shredded
5 fresh chillies, chopped
1 t sugar
1 t white wine
1 t sesame oil
1 t vinegar
1 cup soy sauce

Combine all ingredients and stir well. Use as a dip for a variety of seafood dishes.

# *Pickles and Curries*

# Pete Britz's pickled mussels

3 cups black mussels
1 cup white vinegar
1 t whole allspice
4 bay leaves

Boil mussels until shells open, discard shells and remove 'heads'. Simmer vinegar, allspice and bay leaves gently for half an hour. Place mussels in jars and pour vinegar mixture over. Seal when cool and refrigerate. (Best served after two weeks.)

---

# Tom Pike's pickled fish

firm fish, cut into serving-sized portions
seasoned flour to coat
2 onions, sliced
1 t salt
1 T sugar
2 t curry powder
½ t turmeric
1 cup vinegar
2 T cornflour to thicken

Prepare as for curried fish, but do not add the water. Coat the fish with seasoned flour only, without egg, and fry. Bottle while mixture is hot and seal with a tightly-fitting lid. Leave for about one month, mainly to soften the bones. Serve with salads.

---

# *Denise Oosthuizen's pickles*

1. Black and white mussels or freshwater mussels
2. Periwinkles
3. Roe
4. Perlemoen (mother-of-pearl)
5. limpets

sugar
vinegar
mixed pickle spices

Take equal volumes of sugar and vinegar, 2 cups of water per litre of vinegar and 1 packet of mixed pickle spices per litre and boil for 30 minutes.

*1 and 2:* Steam mussels or periwinkles until easily removable from shell, clean and put into a well-sealed bottle. Pour the boiling mixture into the bottle until it overflows, making sure there are no air bubbles in the bottle. Put the lid on and close tightly. Leave for at least one week before use.

*3:* Boil roe in salt water for 5 minutes and follow the same procedure as for 1 and 2.

*4 and 5:* Cut perlemoen or limpets into small blocks of cocktail snack size and boil in salt water for 5–8 minutes; then follow the same procedure as for 1 and 2.

Serve at cocktail parties or on crackers as snacks. Can be used in seafood salads, using oil and lemon juice as a dressing (a little sugar can be added).

## *Tom Pike's curried fish*

firm fish, cut into serving-sized portions
2 T flour *or* batter
2 onions, sliced
1 t salt
1 T sugar
2 t curry powder
½ t turmeric
1 cup vinegar
1 cup water
2 T cornflour to thicken

Coat pieces of fish with seasoned flour or batter, then fry. Keep aside. Using another saucepan, fry onions but do not brown. Add salt, sugar, curry powder and turmeric. Add vinegar and water. Thicken with cornflour and bring to the boil. Pour over the fried fish and serve. (Enough sauce for 4 servings.)

---

## *Curried fish*

2 kg firm fish such as kabeljou, kingklip *or* hake
4 medium-sized onions, sliced
3 cups vinegar
1 cup water
½ t cayenne pepper
3–4 t salt
½ cup sugar
2 T apricot jam
2 T curry
1 T turmeric
½ cup raisins *or* sultanas
6 lemons *or* 3 bay leaves
3 T cornflour

Fry the fish in the usual way. Boil onion until tender. Drain. Mix the onion, sugar, curry powder, turmeric, salt, cayenne pepper, vinegar, apricot jam and water in a saucepan, add lemon or bay leaves and cook for another 5 minutes. Mix the cornflour with a little water, add to saucepan and cook for another 10 minutes until thick. Add the raisins or sultanas. Pour over the fish and allow to cool. Ready to serve in 2–3 days' time.

---

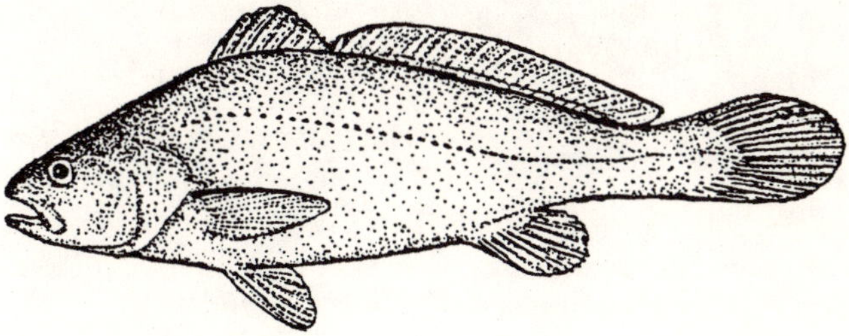

# *Fish curry*

500 g frozen fish, cut into bite-sized pieces
¼ cup oil
½ cup onion, chopped
½ t garlic, minced
1 T curry powder
⅛ cup celery leaves
½ cup tomato, chopped
¾ cup vegetable bouillon (stock)
½ cup water
1–1½ T lemon juice
1 t lemon peel
1½ apples, peeled, cored and chopped
¼ cup chutney
1 t salt
¼ t pepper

Sauté onion, garlic and curry powder in oil for about 5 minutes. Add celery leaves, tomato, bouillon and water. Bring to the boil. Add fish, cover and simmer for half an hour. Stir in lemon juice, lemon peel, apple and chutney. Season with salt and pepper. Simmer uncovered for half an hour. Serve with rice, coconut, raisins, chutney, chopped scallions and chopped nuts. (Serves 3–4.)

# *Braais and Potjiekos*

# BRAAIS

## *Louis Botha's silverfish braai*

Braai a medium-sized, fresh whole silverfish (gutted but not de-scaled) over an open fire of medium heat. Turn the fish around carefully every 5 minutes. Bake each side for about 10–15 minutes. Remove from fire. Gently remove the head and tail and carefully flap the two halves open from the ventral side. Spread open in plate. Eat with fingers. Then you have tasted fish!

---

## *Tom Hecht's 'galjoen-op-die-rooster'*

Catch a galjoen (white water is a prerequisite). Do not scale the fish. 'Fleck' the fish from the top through the head and be careful not to cut through the abdominal wall. Remove the guts, place the fish on the 'rooster' (grid) scale-side down, and dab on pats of butter. Add salt and pepper to taste. Place 'rooster' on the coals. When the fish is nearly done, turn it flesh side down for a short while and turn again. Add more butter pats and wait for these to melt. Serve and enjoy.

---

## *Denise Oosthuizen's fish and greenpepper braai*

500 g fish, cubed
tinfoil
6 green peppers
2 T lemon juice
black pepper to taste
1 t garlic salt
2 onions, chopped
1 t fish spices
2 T mayonnaise
2 boiled potatoes, cubed

Cut the tops off the green peppers and retain to use as lids later on. Clean out green peppers without piercing the sides. Season fish, potato and onion with fish spices, black pepper, garlic salt, lemon juice and mayonnaise to taste (a little sugar can be added). Mix well. Stuff the green peppers with the mixture, put lid back on and wrap in tinfoil. Braai in an upright position on top of the grill over medium coals until tender.

---

## *Mike Coke's barracuda braai*

thick crosscuts (steaks) of barracuda *or* any firm-fleshed fish
plenty of mayonnaise to cover

Cover fish with mayonnaise on the upper side only. Grill or braai slowly, allowing the sauce to penetrate the fish deeply.

---

# *Snoek braai*

1 medium-sized mild-cured snoek, cut open
pepper to taste
*or*
1 medium-sized fresh snoek, cut open
salt and pepper to taste

## Basting mixture

juice of 3 lemons
juice of 1 orange
250 g margarine, melted

Put open snoek on 'rooster' and place over hot coals with skin side facing coals. Pour basting mixture on open side and leave until bubbles are visible. Turn 'rooster' and brown open side. Turn 'rooster' and keep pouring mixture on open side until cooked.

---

# *Peter Britz's fish braai baste*

*This baste can be used for all braai fish such as snoek, geelbek, kob, musselcracker, 'poenskop' and galjoen.*

a generous handful of chopped fennel
juice of 2–3 lemons
2 t brown sugar
fresh coriander leaves (optional)
200 m*l* oil
200 m*l* white wine
1 t salt

Blend ingredients thoroughly and use baste liberally at your next fish braai.

---

# *Traditional fish braai*

2 kg fresh fish, cleaned and washed

## Marinade

150 m*l* dry white wine
juice and rind of half a lemon
1 onion, thinly sliced
1 carrot, scraped and thickly sliced
10 m*l* chopped parsley
1 celery stalk, sliced
3 m*l* dried thyme
1 bay leaf
37,5 m*l* oil
freshly ground black pepper

Combine all ingredients for marinade. Place fish on large double sheet of tinfoil, shiny side inside. Crumple up the sides slightly and pour in marinade. Marinate in refrigerator for 3–4 hours. Place the open foil parcel on the braai grid and cook slowly, basting frequently for about one hour or until fish is tender. Serve with lemon wedges. (Serves 6.)

---

## *Allan Heydorn's perlemoen potjie*

6 medium-sized perlemoen
1 pkt seafood rice
2 T margarine
1 cup diced potatoes *and/or* 1 cup mushrooms
meat tenderiser
1 *l* hot water

Tenderise perlemoen by bashing it with the broad side of a bottle. Cut into pieces. Add meat tenderiser and allow to stand for at least 1 hour. Heat black cast-iron pot. Sauté perlemoen in margarine. Add rice, potatoes and water. Boil for 20 minutes. Do not overcook perlemoen as this toughens it. Serve immediately.

## *Denise Oosthuizen's perlemoen potjie*

5 large perlemoen, cleaned and cut into strips small enough to fit into mincer or food processor
1 tin creamed mushrooms (285 g)
3 large onions
black pepper
1 pkt rice *or* seafood rice
5 m*l* vinegar

Boil perlemoen strips in sea water or salt water with vinegar for 10 minutes. Mince or process the perlemoen and onions. Put into a black pot with mushrooms and 2 cups of water, adding black pepper to taste. Bring to the boil over very low heat and boil for 6–10 minutes. Serve on rice or seafood rice, prepared in a separate pot.

# *Denise Oosthuizen's crayfish potjie*

5 crayfish
butter
3 large onions
garlic
black pepper
5 m*l* vinegar
1 cup seafood mayonnaise
1 pkt rice *or* seafood rice

Add a little water and some vinegar to a black pot, bring to the boil and steam crayfish for about 12 minutes. Leave to cool, clean and break crayfish into small pieces, keeping the unshelled legs separate. Fry onions, plenty of garlic and rice in butter over medium heat until golden brown. Add water and cook rice over low heat for about 20 minutes. Remove pot from direct heat and add crayfish pieces and black pepper to taste. Mix well and add seafood mayonnaise until a creamy texture is obtained. Serve with crayfish legs and slices of lemon.

# *Denise Oosthuizen's periwinkle potjie*

periwinkles (*or* 'alikreukel')
Tabasco sauce
peri-peri sauce
1 tin tomato paste (200 g)
onion and garlic salt
vinegar
1 bottle Italian salad dressing
rice *or* seafood rice
2 cups water
hairpins to remove small periwinkles from shell

Mix water, tomato paste and salad dressing in a black pot, add a dash of vinegar, Tabasco sauce, peri-peri, onion and garlic salt to taste, and bring to the boil over low heat. Add the periwinkles and boil for another 10–15 minutes. Serve with rice or seafood rice.

# *Denise Oosthuizen's octopus potjie*

1 octopus
2 onions, chopped
3 tomatoes, chopped
2 potatoes, chopped
1 cup black mussels *or* fish, chopped
salt and black pepper
1 T vinegar
2 T butter
parsley
paper clips or thin wire to close end of tentacles

Cut off tentacles and clean well, leaving one end closed. Chop rest of edible octopus into very small pieces. Add onion, potato, tomato and mussels or fish. Season with black pepper, salt, vinegar and parsley to taste. Mix well and stuff tentacles, then close the open end with wire or a paper clip. Fry in butter over low heat in a black pot for 10–15 minutes or until golden brown. Stuffed tentacles can also be wrapped in tinfoil and braaied over the coals.

---

## *Denise Oosthuizen's mussel potjie*

2 cups black *or* freshwater mussels
black pepper
2 t Tabasco sauce
1 bottle Italian salad dressing
3 large onions, chopped
3 large cloves of garlic, chopped
500 g spaghetti *or* rice
2 T butter
about 100 m*l* water

Put water in a black pot and bring to the boil. Add mussels and steam until shells open and mussels can be removed easily. Discard shells and clean mussels. Fry onion and garlic in butter over low heat until golden brown. Add mussels and salad dressing. Season with black pepper and Tabasco sauce. Boil for 6–10 minutes over low heat. Serve on spaghetti or rice prepared in a separate pot.

---

# *Denise Oosthuizen's roe potjie*

1 cup roe
1 t garlic salt
½ t pepper
1 T lemon juice
2 T butter

Roll roe in a mixture of garlic salt and pepper and fry in butter over medium heat in a black pot for 5–8 minutes. Serve with lemon juice.

# *Microwave Recipes*

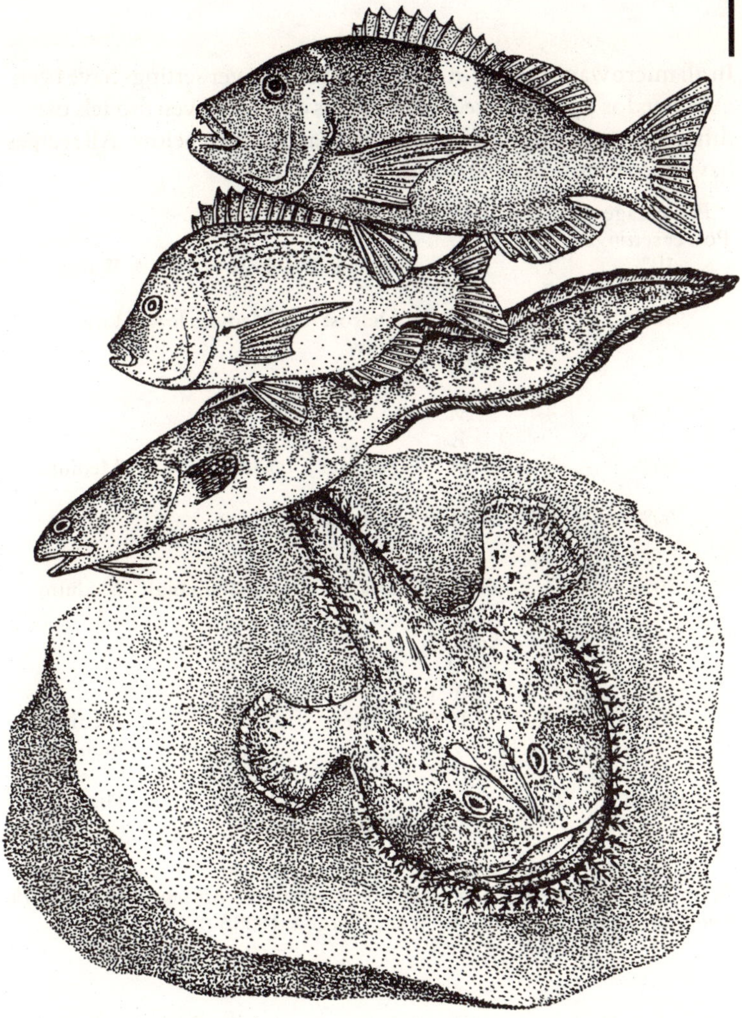

# *Microwave power settings*

In all microwave recipes in this section the power settings have been expressed as a percentage. However, as different oven models use different terminology, a conversion chart is given below. All recipes have been tested on a 650 watt microwave oven.

| Percentage Power settings | | | | | | | |
|---|---|---|---|---|---|---|---|
| 10% | 1 | | L | | Warm | Low | Warm |
| 20% | 2 | | O | | | | |
| 30% | 3 | | W | | Defrost | Medium low | Low |
| | | | | | Stew | | |
| | | | | | Braise | | |
| 40% | 4 | | M | | | | |
| | | | E | | | | |
| 50% | 5 | | D | | Simmer | Medium | Medium |
| | | | I | | | | low |
| 60% | 6 | | U | | Defrost | | |
| | | | M | | Bake | | |
| 70% | 7 | H | N | F | Roast | Medium high | Medium |
| | | I | O | U | | | |
| | | G | R | L | | | |
| 80% | 8 | H | M | | Reheat | | |
| | | | A | P | | | |
| 90% | 9 | | L | O | Sauté | | Medium |
| | | | | W | | | high |
| | | | | E | | | |
| 100% | 10 | | | R | High | High | High |
| | | | | | Normal | | |
| | | | | | Full power | | |

(With acknowledgement to Christine Cashmore. *Microwave Cooking the South African Way,* J.P. van der Walt, Pretoria, 1988.)

# *Haddock pâté*

500 g haddock fillets, cut into even portions
1 small onion, finely chopped
250 g cream cheese
10 m*l* lemon juice
salt and black pepper to taste

Arrange fish in a casserole dish with the thickest portions to the outside, place onion on top, cover and microwave on 100% for 5 minutes. Allow to cool, then liquidise the fish with the rest of the ingredients until well blended. Adjust seasoning. Fill ramekin dishes with mixture and refrigerate until firm. Garnish with lemon wedges and parsley sprigs. Serve with melba toast.

# *Baked haddock and potato*

500 g cooked, flaked haddock
4 potatoes, peeled
1 tin of condensed mushroom soup (410 g)
125 m*l* milk
5 m*l* prepared mustard
12,5 m*l* chopped parsley
30 g grated cheese
paprika
tomato to garnish

Boil potatoes for about 8 minutes. Cool and slice. Mix soup with milk, mustard and parsley. Layer the potato, haddock and soup mixture in a casserole dish. Repeat and cover with slices of tomato. Microwave on 50% for about 10 minutes. Sprinkle with cheese and paprika and serve.

# *Haddock pie*

4 fillets of haddock (400 g)
3 potatoes
milk
onion salt
2 eggs
30 m*l* chopped parsley
crushed cornflakes

Microwave potatoes on 100% for 8–10 minutes. Allow to stand for 5 minutes. Place haddock fillets on a plate. Cover with vented plastic wrap and microwave on 100% for 5–6 minutes. Peel potatoes and mash, adding a little milk, onion salt, eggs and parsley. Flake haddock and add to the potato. Put in a casserole dish and sprinkle with cornflake crumbs. Microwave on 50% for 6–9 minutes.

---

# *Fish kedgeree*

250 g smoked haddock
2 hard-boiled eggs
25 g (30 m*l*) margarine
175 g (225 m*l*) cooked rice
30 m*l* cream
salt and cayenne pepper
15 m*l* parsley, chopped

Put the haddock in a casserole dish, cover and microwave on 100% for about 4 minutes or until cooked. Allow to stand for 4 minutes, skin and flake. Chop one egg and slice the other into rings. Microwave the margarine on 100% power for 2 minutes or until melted. Add the fish, chopped egg and rice. Microwave on 100% for about 3 minutes or until heated through. Stir once during cooking. Gently mix the cream, salt and cayenne pepper and garnish with egg slices and parsley.

---

# Haddock pie

500 g haddock
4 medium-sized potatoes
1 tin cream of mushroom soup (410 g)
750 m*l* milk
5 m*l* prepared mustard
1 T chopped parsley
30 g grated cheese
paprika

Prick potatoes and microwave on 100% for 7 minutes. Cover the haddock with 125 m*l* of the milk, cover dish and microwave on 100% for 5 minutes. Mix soup with the rest of the milk, mustard and parsley. Peel potatoes and slice. Flake haddock and then layer potato, haddock and soup mixture in a glass casserole dish, repeating layers and ending with grated cheese. Sprinkle with paprika and parsley and microwave on 50% for 10 minutes.

# Fish pie

400 g cooked, flaked haddock *or* 1 tin pilchards in tomato sauce (425 g), flaked
1 pkt instant mashed potato (112 g)
1 small onion, diced
fennel *or* parsley
1 t dried basil
salt and pepper
2 eggs, beaten
250 m*l* boiling water

Prepare mashed potato with boiling water according to the instructions on packet and put in a Pyrex dish. Add herbs, onion and seasoning. Blend in pilchards or haddock. Add eggs and beat in well. Microwave on 100% for 12–15 minutes. Garnish with fennel or parsley.

# *Fish cakes*

500 g haddock
300 g smoked haddock
100 m*l* milk
salt and pepper
1 chopped onion
boiling water
½ pkt instant mashed potato (56 g)
2 m*l* thyme
50 m*l* chopped parsley
cake flour
30 m*l* oil

Put fish, milk, seasonings and onion in a large bowl, cover with vented plastic and microwave on 100% for 9–10 minutes. Drain off liquid and remove skin and bones from fish. Process fish and onion until smooth. In a separate bowl, combine instant potato, fish liquid and sufficient boiling water to form a stiff consistency. Add this mixture, thyme and parsley to the fish. Process and shape into fish cakes. Sprinkle both sides with a little flour. Preheat browning dish for 8 minutes. Add 15 m*l* oil and microwave at full power for 1 minute. Add half the fish cakes, microwave at full power for 2 minutes, then turn and microwave at full power for another 2 minutes. Drain on paper towels. Wipe the browning dish with paper towels and microwave at full power for 3 minutes to reheat. Repeat procedure with the remaining fish cakes. Serve hot, or cold in children's lunch boxes.

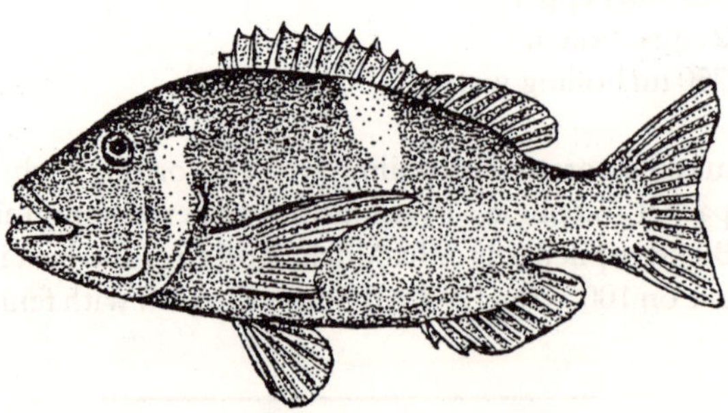

# *Passover gefilte fish*

*This is a traditional Jewish dish.*

750 g hake
2 carrots, sliced
1 medium-sized onion
1 stick celery
30 m*l* matzo meal *or* 30 m*l* white breadcrumbs
salt and black pepper
15 m*l* oil
3 eggs
Fondor *or* Aromat
500 m*l* chicken *or* fish stock – see below
5 m*l* yellow food colouring
15 m*l* gelatine
parsley and sliced carrot to garnish

Process fish, carrots, onion, celery and parsley. Mix matzo meal, seasoning, oil and egg well and arrange in a single layer in a large casserole dish. Add stock, food colouring and gelatine. Cover and microwave on 70% for 20 minutes. Uncover and allow to cool. Refrigerate until well chilled and jelled. Cook carrot in 5 m*l* water on 100% for 2 minutes. Arrange fish balls on a flat serving platter with the jellied liquid and garnish each fish ball with a slice of carrot and a little parsley. (Serves 12.)

## Fish stock

heads and bones of hake
2 carrots
1 onion
celery
salt and pepper to taste

Put fish heads and bones, carrots, onion, celery, salt and pepper in a large dish and microwave on 100% for 1 hour, and then on 70% for 30 minutes. Strain and set aside.

# *Gingered hake with lemon*

900 g fillet of hake, skinned and cut into serving portions
15 g fresh ginger root, peeled and grated *or* 2 t powdered ginger
grated rind and juice of 1 lemon
3 T oil
1 t ground turmeric
½ t ground coriander
¼ t cayenne pepper

Put fish in a shallow dish. Mix the remaining ingredients and spoon over the hake. Leave for 2 hours at room temperature to marinate. Cover and microwave on 100% for 8 minutes. Leave to stand for 4 minutes before serving.

---

# *Bacon bake*

600 g hake *or* kingklip, skinned
125 g bacon
2 onions, sliced
12,5 m*l* margarine
salt and pepper *or* Aromat
150 m*l* tomato sauce *or* 1 tin tomato paste (112 g)
150 m*l* water
5 m*l* sugar
5 m*l* Worcestershire sauce
1 m*l* cayenne pepper
chopped parsley to garnish

Microwave bacon on a ridged microwave dish or paper plate covered with grease-proof paper on 100% for about 2 minutes and crumble. Sauté onion and margarine in a bowl on 100% for 4 minutes, stirring halfway through. Put fish in a large buttered casserole dish. Sprinkle with salt, pepper, sugar and Aromat, then cover with chopped bacon and onion. Mix tomato sauce, water, Worcestershire sauce, cayenne pepper and a pinch of salt. Pour sauce over fish, carefully covering the complete surface. Microwave on 100% for about 10 minutes. Allow to stand. If the fish flakes easily it is cooked. Serve with creamy mashed potatoes and peas.

---

# *Gherkin and sandwich spread fish bake*

750 g hake *or* kingklip
100 m*l* white wine
30 m*l* sandwich spread
50 m*l* cream
75 m*l* cornflake crumbs *or* breadcrumbs
2,5 m*l* salt
1 bay leaf
125 m*l* mayonnaise
50 m*l* chopped gherkins
150 m*l* grated cheese
paprika
a few peppercorns

Put fish in a casserole dish with peppercorns, wine and bay leaf. Cover with vented plastic wrap and poach on 100% for 5 minutes. Pour off liquid and allow to stand for 3 minutes. Mix mayonnaise, gherkins and sandwich spread, pour over the fish and sprinkle with crumbs. Microwave on 100% for 2 minutes, sprinkle with grated cheese and paprika and microwave on 100% for another 3 minutes or until cheese has melted. Garnish with tomato slices and parsley.

---

# *Kingklip with yoghurt*

400 g kingklip *or* hake
salt and pepper
1 grated onion *or* 5 m*l* onion flakes
20 m*l* grated Parmesan cheese
100 m*l* plain yoghurt
25 m*l* mayonnaise *or* seafood mayonnaise
chopped parsley

Season fish with salt and pepper and put in a greased dish. Mix the onion, Parmesan cheese, yoghurt and mayonnaise. Spread over the fish. Sprinkle with chopped parsley. Cover with vented plastic wrap and bake on 100% for about 4 minutes. Allow to stand. If the fish flakes easily with a fork it is cooked. If not, return for another minute. Garnish with lemon slices and serve with a baked potato.

---

# *Seafood cocktail*

750 g monkfish, stockfish, hake *or* tuna
100 m*l* dry white wine
1 bay leaf
1 small onion, sliced
2 sprigs parsley

## Sauce

125 g smooth cottage cheese
2 m*l* ginger
2 m*l* curry powder
2 m*l* sugar
1 T sherry
1 T tomato sauce
5 m*l* Worcestershire sauce

Put fish in a microwave-proof dish, cover with onion, parsley, bay leaf and pour over the wine. Cover with vented plastic wrap and microwave on 100% for about 7–8 minutes. Allow to stand, drain, season to taste and refrigerate. Whisk all the sauce ingredients together and refrigerate. Cube the chilled fish, spoon into cocktail glasses and top with a generous helping of sauce. Dust with paprika, garnish with a sprig of parsley and serve very cold. This sauce can be used on special occasions with prawns or crayfish.

---

# *Pasta shells in tuna sauce*

2 × 200 g tins tuna, drained
250 g pasta shells
40 g margarine
100 g button mushrooms, sliced
40 g flour
600 m*l* milk
salt and freshly ground black pepper
4 hard-boiled eggs, quartered
100 g grated cheddar cheese
finely chopped parsley

Cook the pasta shells in a large saucepan with boiling, salted water until tender. Drain and rinse well. Heat the margarine in a microwave-proof bowl on 100% for 1 minute. Stir in the mushrooms and microwave on 100% for 1 minute. Stir in the flour and gradually blend in the milk. Microwave on 100% for 6 minutes, stirring once during cooking and once at the end. Add the seasoning, tuna, eggs and pasta and coat evenly with the sauce. Put the mixture in a microwave-proof serving dish. Cover with cheese and microwave on 100% for 10 minutes or until heated through. Brown under the oven grill until the cheese is golden. Garnish with parsley. (Serves 4.)

---

# *Lynn's tuna lasagne*

2 × 200 g tins tuna, drained and flaked
3/4 pkt green noodles (375 – 400 g)
12,5 m*l* oil
1 large onion, finely chopped
5 m*l* salt
2 cloves garlic, crushed *or* finely chopped
1 tin tomato soup *or* 1 pkt tomato soup mix combined with 250 m*l* water
salt, pepper and sugar to taste
12,5 m*l* margarine

## Sauce

250 g smooth cottage cheese
250 m*l* cream
250 m*l* grated cheese

Blend sauce in a separate bowl.

Boil noodles, oil and salt in water for about 10 minutes or until tender. Set aside. Melt margarine for 30 seconds. Add chopped onion and garlic. Microwave on 100% for about 1½ minutes. Add tuna, tomato soup, salt, pepper and sugar. Simmer on 70% for about 10 minutes.

Layer the noodles, tuna mixture and cheese mixture alternately in a greased casserole dish. Repeat, ending with the sauce. Sprinkle with grated cheese and paprika. Bake on 50% for about 12 minutes. To brown and crisp the cheese, place under the grill for a few minutes.

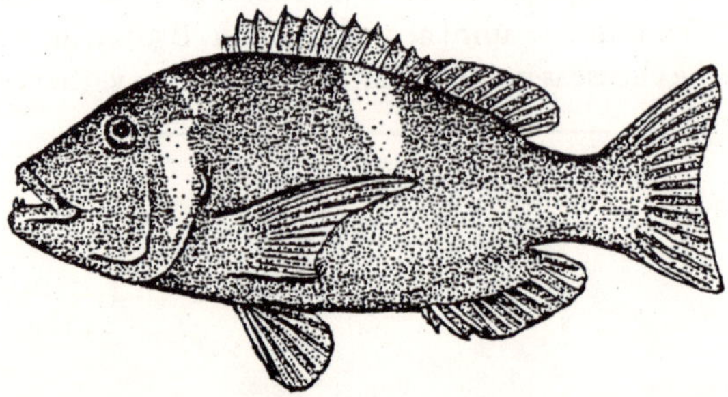

# *Savoury tart*

1 tin tuna (200 g), drained and flaked
1 pkt cheddar biscuits, crushed
75 g margarine
1 onion, chopped
1 green pepper, chopped
2 sticks celery, chopped
2 eggs
1 tin ideal milk (170 g)
herbs and spices
125 m*l* cheese
10 m*l* prepared mustard
5 m*l* Aromat
2 m*l* pepper

Melt margarine on 100% for 30 seconds, add to biscuits and press into a pie dish. Beat eggs, milk, herbs and spices. Sauté ½ teaspoon margarine, onion, celery and green pepper on 100% for 3 minutes and combine with tuna and egg mixture. Microwave on 70% for 5–6 minutes until slightly set and partially cooked. Stir a few times during the cooking process. Spoon onto a pie crust, sprinkle with cheese and cook on 50% for 5 minutes.

---

# *Cheesy salmon or tuna bake*

1 tin salmon *or* tuna (200 g), flaked
250 m*l* cooked rice
125 m*l* mayonnaise
250 m*l* milk
1 egg, lightly beaten
12,5 m*l* chopped parsley
2,5 m*l* salt
pinch of pepper
125–250 m*l* grated cheese

Combine all ingredients. Pour into a shaped, greased mould. Microwave on 50% for 15–20 minutes. Serve with baked potatoes and a green salad.

# *Quick tuna chowder*

1 tin tuna (200 g), drained and flaked
30 m*l* oil
8 spring onions, chopped
1 m*l* each dried origanum and basil
1 carrot, grated
1 potato, grated
500 m*l* boiling water
1 chicken stock cube
100 g smooth cottage cheese
100 m*l* milk
salt and freshly ground black pepper
Tabasco sauce
10–15 m*l* lemon juice

Microwave oil in a deep casserole dish on 100% for 45 seconds. Add most of the spring onion, reserving a little of the green portion to garnish. Stir to coat with oil and microwave for a further 3 minutes. Add the tuna, herbs, vegetables, water and stock cube. Cover and micro-wave on 100% for 10 minutes. Blend the cottage cheese and milk until smooth and stir into the soup. Season with salt, pepper, Tabasco sauce and lemon juice. Microwave on 100% for 2–3 minutes to reheat. Spoon into bowls, and sprinkle with spring onion. (Serves 4–6.)

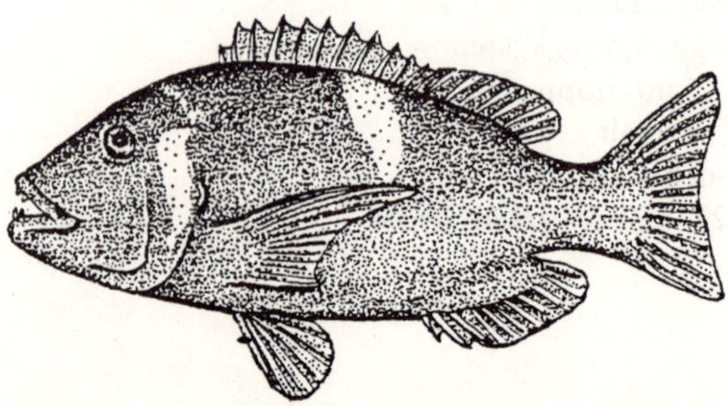

# *Paella*

1 tin mussels *or* smoked oysters (105 g)
250 g shelled shrimps *or* prawns (optional)
500 g cubed, cooked chicken
1 red pepper, sliced (optional)
1 green pepper, sliced
1 onion, sliced
1 clove garlic, crushed
250 m*l* mushrooms, sliced
62,5 m*l* oil
375 m*l* uncooked rice
750 m*l* hot chicken stock *or* 650 m*l* stock and 100 m*l* white wine
5 m*l* salt *or* Aromat
5 m*l* turmeric
5 m*l* marjoram
250 m*l* frozen peas
stuffed olives (optional)

Set aside a few mussels, shrimps, prawns, olives and a sprig of parsley for garnishing.

Heat oil in a large casserole or browning dish on 100% for 2 minutes. Sauté green and red peppers, onion, garlic and mushrooms on 100% for about 2 minutes. Add the rice, turmeric, marjoram, salt or Aromat, pepper and stock. Cook on 50% for about 25 minutes. Add chicken, mussels or oysters, shrimps or prawns and frozen peas. Mix well. Cook on 50% for another 7 minutes. Toss in 12,5 m*l* butter. Garnish with the shrimps, mussels, prawns and olives and sprinkle with chopped parsley. (Serves 4–6.)

*Two variations to this recipe are to add 125 g cooked, chopped bacon or 125 g cubed ham instead of the mussels or prawns.*

# *Snoek pie*

## Crust

75 g margarine
150 g flour
1 egg yolk
60 m*l* water

## Filling

265 g snoek, rinsed *or* soaked to remove saltiness, and flaked
50 g margarine
50 g flour
250 m*l* asparagus salad cuts, drained
450 m*l* milk and drained asparagus liquid
100 g grated cheddar cheese
1 egg, beaten
salt and freshly ground black pepper
15 m*l* chopped parsley
breadcrumbs

Process or rub the margarine into the flour. Add the egg yolk and water to form a dough. Roll out on a floured surface and line a lightly-greased pie dish. Prick and bake blind on 100% for 4–5 minutes. Meanwhile, microwave 50 g margarine on 100% for about 2 minutes or until melted. Add the flour and mix until smooth. Gradually stir in the liquid. Microwave on 100% for about 6 minutes until sauce thickens. Stir every 2 minutes. Stir in the snoek, asparagus, cheese and egg. Season with salt and pepper and add the parsley. Pour the filling into the pie crust and microwave on 70% for 10 minutes. Top with grated cheese and breadcrumbs and microwave on 100% for another 4 minutes. Allow to stand for about 10 minutes until set.

# *Kipper and lemon pâté*

1 pkt boneless kipper fillets (200 g)
50 g butter
30 m*l* chopped onion
5 m*l* grated lemon peel
5 m*l* lemon juice
salt and freshly ground black pepper
45 m*l* cream
lemon leaves for garnishing
10 m*l* chopped chives
300 m*l* boiling water

Pierce the kipper packet and place in a microwave-proof dish. Add water. Microwave on 100% for 5 minutes. Cut open, drain the fish, skin and liquidise. Microwave the butter on 100% for 1 minute and stir in the onion. Microwave on 100% for 1 minute. Add the butter, onion, lemon peel, lemon juice, black pepper, salt and chives to the fish. Liquidise the mixture, add the cream and season. Spoon the pâté into individual dishes and refrigerate. Garnish and serve on hot toast.

---

# *Baked fish with Parmesan*

500 g fish fillets
60 m*l* Parmesan cheese
2 egg whites, lightly beaten
200 m*l* seasoned breadcrumbs
3 m*l* garlic salt

Rinse the fillets and pat dry. In a small bowl, combine crumbs, cheese and garlic salt. Dip fillets in egg white and then in the crumb mixture, coating well. Arrange fillets in a shallow dish, turning the tails under to prevent overcooking. Cover with a lid or vented plastic wrap. Microwave on 100% for 6–7 minutes. Allow to stand for 2 minutes. Serve with lemon wedges.

---

# *Fish in white wine*

1 kg frozen line fish such as red roman, dageraad and slinger
1 onion, chopped
½ green pepper, diced
4 T butter
1 tin sliced mushrooms (285 g)
2 T lemon juice
2 T freshly chopped parsley
1 clove garlic, crushed
½ t thyme
1 tin prawns (100 g) (optional)
pinch of salt
dash of Tabasco sauce
60 m*l* white wine

Microwave fish on 100% for a few minutes and reserve. Put remaining ingredients, except prawns and wine, in an ovenproof dish, cover with vented plastic and microwave on 100% until vegetables are tender. Stir in wine and prawns, add fish, cover with sauce and microwave on 100% for another 4 minutes or until fish is cooked.

---

# *Baked fish*

500 g kingklip *or* stockfish
1 t sugar
60 m*l* dry white wine
200 m*l* breadcrumbs (about 2 thick slices of bread)
margarine
2 tomatoes, liquidised
1 small onion, chopped
1 t garlic salt
2 T finely chopped parsley
½ t mixed herbs
salt and pepper

Season fish and put in a greased microwave-proof baking dish. Cover with wine, tomatoes, onion and sugar. Mix the breadcrumbs, garlic salt, mixed herbs, parsley, salt and pepper. Sprinkle over the fish. Microwave on 100% for 10 minutes. Allow to stand for 5 minutes before serving.

---

# *Fish bobotie*

500 g cooked, flaked white fish
15 m*l* margarine
1 onion, chopped
1 slice white bread
125 m*l* milk
3 eggs
20 m*l* lemon juice
10 m*l* curry powder
2,5 m*l* nutmeg
2,5 m*l* turmeric
25 g seedless raisins
60 m*l* chutney
salt and black pepper
lemon *or* bay leaves for garnishing

Microwave the margarine on 100% until melted. Stir in the onion and microwave on 100% for 3 minutes. Stir in the fish. Soak the bread in a mixture of 60 m*l* milk and one egg in a separate container. Add the lemon juice, spices, chutney, salt and pepper and mix well. Combine the curry mixture with the fish and spoon into a shallow greased dish. Beat the 2 eggs with the remaining 65 m*l* milk. Pour the egg mixture over the curry. Garnish with lemon or bay leaves. Microwave on 50% for about 25 minutes or until bobotie has set.

---

# *Fish in beer*

675 g fish fillets, skinned and cut into even-sized pieces
juice of 1 lemon
salt and freshly ground black pepper
150 m*l* beer
25 g margarine
1 large onion, sliced
2 T wholewheat flour
4 T chopped parsley

Put fish in a shallow dish, sprinkle with lemon juice and season. Leave for 1 hour at room temperature. Pour the beer over the fish and microwave on 100% for 6 minutes. Microwave the margarine and onion on 100% for 3 minutes in a bowl covered with vented plastic wrap. Stir in the flour and the liquid from the cooked fish. Microwave uncovered on 100% for 4 minutes, stirring after 2 minutes and again after 3 minutes. Stir in the parsley. Pour the sauce over the fish and microwave on 100% for 1 minute. Leave to stand for 4 minutes before serving.

# *Mushroom lemon fillets*

400 g frozen fish fillets, thawed
100 g butter *or* margarine
250 m*l* sliced fresh mushrooms *or* 285 g tinned mushrooms
(if tinned mushrooms are used, drain and reserve liquid)
125 m*l* sliced onion
15 m*l* flour *or* cornflour
5 m*l* salt
5 m*l* grated lemon peel
2,5 m*l* dried parsley
2,5 m*l* pepper
100 m*l* milk

Melt butter on 100% for 1 minute and microwave onions on 100% for 2 minutes. Add mushrooms and microwave on 100% for 3–5 minutes. Blend flour, salt, parsley, lemon peel and pepper. Stir in milk and mix well. Microwave on 100% for 5–6 minutes until sauce thickens, stirring twice. Arrange fillets with the thick edges towards the outside of dish and spoon sauce over. Cover and microwave on 50% for 5–8 minutes or until fish flakes easily. Allow to stand and serve with grated lemon rind and chopped parsley.

# *Pickled fish*

1 kg white fish fillets
salt and pepper
200 m*l* vinegar
125 m*l* dry white wine
2 onions, sliced
15 g (15 m*l*) margarine
10 m*l* sugar
10 m*l* curry powder
2 T apricot jam
25 m*l* chutney
3 m*l* allspice
5 m*l* turmeric
2,5 m*l* salt
5 m*l* cornflour

Put fish in a large dish, season with salt and pepper and cover. Microwave on 100% for 8 minutes. Mix all the remaining ingredients, except cornflour, and microwave on 70% for 10 minutes. Mix cornflour with water to make a smooth paste. Add to the curry mixture. Microwave on 100% for about 5 minutes or until the sauce thickens. Layer fish and sauce alternately in a serving dish. Cover once the dish has cooled. Allow to marinate in the refrigerator for a few days before serving.

# Cucumber fish bake

600 g fish, flaked
½ English cucumber, sliced
1 pkt white onion soup
1 cup plain yoghurt

Layer fish and cucumber in a microwave-proof casserole dish. Blend yoghurt and soup powder and pour over fish. Microwave on 70% for 8 minutes.

---

# Curried fish

1 kg kabeljou *or* kingklip *or* hake
2 green peppers, cut into strips
250 m*l* chopped tomato
10 m*l* crushed garlic
125 m*l* oil
1 onion, chopped
pinch of salt
½ t mustard seed
1 T lemon juice
2 T curry powder (Masala)
½ bunch chopped dhania (available from Indian grocers) *or*
2 t ground coriander

Wash fish and pat dry. Rub 1 tablespoon of curry powder over fish and marinate in the lemon juice for 30 minutes. Heat a browning dish on 100% for 6 minutes. Add half the oil and heat on 100% for 1 minute. Fry the fish for 1 minute on each side. Pour the remaining oil in the dish and heat for 30 seconds on 100%. Fry the peppers, onion, mustard seed and garlic on 100% for 4 minutes. Stir in the curry powder, tomato, coriander and salt. Arrange the fish in the sauce, cover and microwave on 100% for 8 minutes. Garnish with the dhania and serve on yellow rice.

---

# *Cheese and parsley sauce*

25 g butter
1 onion, finely chopped
25 g flour
250 m*l* milk
75 g grated cheddar cheese
30 m*l* finely chopped parsley
salt and pepper

Put butter and onion in a microwave-proof jug. Microwave on 100% for 2–3 minutes or until tender. Stir in the flour and blend in the milk, stirring until smooth. Microwave on 100% for 5 minutes, stirring 2 or 3 times during cooking. Add cheese and microwave on 100% for 30 seconds. Add the parsley and seasoning and stir well. (Makes about 250 m*l*.)

---

# *Creamy shrimp sauce*

500 g frozen shrimps, thawed
75 g butter
1 onion, chopped
1 bunch spring onions, chopped
250 m*l* cream
250 g smooth cottage cheese
250 g cottage cheese with chives
15 m*l* chopped parsley
salt and freshly ground black pepper

Put butter, onion and spring onion in a microwave-proof dish. Cover with vented plastic wrap and microwave on 100% for 2 minutes. Add the shrimps and microwave on 100% for another 2 minutes. Add cream and microwave on 100% for 2 minutes. Combine the cottage cheese and gradually stir into prawn mixture. Microwave on 50% for 8 minutes, stir every 2 minutes. Season. Serve with grilled or fried fish and rice. (Serves 6.)

---

# *Saucy fish fingers*

400 g frozen fish fingers

## Tomato and bacon sauce

3 slices bacon
1 tin condensed tomato soup (410 g)
2 T water
60 m*l* chopped celery
2 T chopped onion
½ t origanum
⅛ t garlic powder

Arrange frozen fish fingers in a 3,75 *l* glass baking dish and set aside. Microwave bacon in a ridged dish or on a paper plate on 100% for about 3 minutes or until crisp. Drain and crumble. Stir in remaining sauce ingredients. Place fish on upper rack of oven and sauce on floor of oven. Cook sauce and fish separately if oven has no rack. Microwave both on 100% for 10–14 minutes or until fish and sauce are hot. Stir sauce and allow to stand for 5 minutes. Serve over fish. (Serves 4–5.)

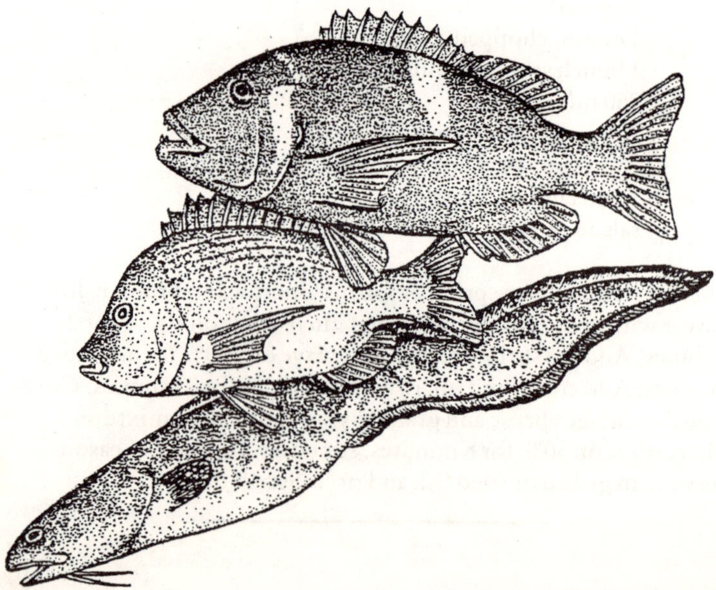

# *Common and scientific names of finfish mentioned in the text*

| | |
|---|---|
| barracuda | *Sphyraena* spp. |
| bass | *Micropterus salmoides* and *M. dolomieu* |
| blacktail | *Diplodus sargus capensis* |
| bream *or* tilapia | Cichlidae |
| carp | *Cyprinus carpio* |
| catfish | Clariidae |
| codfish *or* kabeljou | Moridae and Gadidae (*or see* hake) |
| dageraad | *Chrysoblephus cristiceps* |
| eel (freshwater) | *Anguilla* spp. |
| galjoen | *Coracinus capensis* |
| grunter | *Pomadasys* spp. |
| haddock | *see* hake |
| hake | *Merluccius capensis* |
| kingfish | Carangidae |
| kingklip | *Genypterus capensis* |
| kob | *Argyrosomus hololepidotus* |
| mackerel | *Scomber japonicus* |
| monkfish | Lophiidae (usually *Lophius vomerinus*) |
| mullet | Mugilidae |
| red line fish | Sparidae |
| red snapper | Lutjanidae |
| sea barbel *or* sea catfish | Ariidae |
| snoek | *Thyrsites atun* |
| sole *or* flounder | Soleidae *or* Bothidae |
| tigerfish | *Hydrocynus forskahlii* |
| trout | *Salmo gairdneri* or *S. trutta* |
| tuna | Scombridae *(Thunnus* spp.*)* |
| white musselcracker | *Sparodon durbanensis* |

# *Recommended seafood recipe books*


Antolini, H.L. & D. Van Lear (Eds). (1982). *Sunset Seafood Cookbook*. Lane Publishing Co, Menlo Park, California.

Charlton, W. (Ed). (1981). *The Good Cook – Fish and Shellfish*. Time Life Books, Amsterdam.

David, E. (1955). *A Book of Mediterranean Food*. John Lehmann, London.

Fast, B. (Ed). (1982). *The Catfish Cookbook*. Fast & McMillan Publishers Inc., Charlotte, North Carolina.

Fauny, B.F.M.P. (Translator). (1981). *Sea and Freshwater Fish*. Ferndale Editions, London.

Froud, N. & T. Lo. (1976). *International Fish Dishes*. Sphere Books, London.

Miloradovich, M. (1984). *The Art of Fish Cookery*. Bantam Books, New York.

Pearman, R. & H. Adams (Eds). *More Wild Ways with Cooking*. Wildlife Society, Durban.

Ross, S. (1978). *Fish Cook Book*. Southpoint Publishing, Cape Town.

Shaxson, A., P. Dickson & J. Walker. (Eds). (1978). *The Malawi Cookbook*. Government Printer, Zomba.

Snyman, L. & A. Klarie (1976). *Free from the Sea: The South African Seafood Cookbook*. Don Nelson Publishers, Cape Town.

Snyman L. & A. Klarie (1986). *More Free From the Sea*. Don Nelson Publishers, Cape Town.

Strömsöe, A. (1982). *Fish Book*. Howard Timmins, Cape Town.

Su-Huei, H. (1984). *Chinese Seafood*. Sun-Chuan Publishing Co, Taipei.

Wilkinson, A. (1981). *The Complete South African Fish and Seafood Cookbook*. Central News Agency, Johannesburg.

# *ICHTHOS – for people with an interest in fishes*

## *What is ICHTHOS?*

The world famous ichthyologists Professor J.L.B. Smith and his wife, Professor Margaret M. Smith, established a tremendous rapport with the public and especially the anglers of southern Africa through their scientific publications, popular talks and articles, and radio broadcasts on the subject of fish. They explored our coastline, stopping to talk to fishermen, listening to their tales and offering constructive advice. This intimate association is reflected in the success of the Smiths' various books, especially *Smiths' Sea Fishes*, and in a number of donations that the J.L.B. Smith Institute of Ichthyology receives from anglers, divers, aquarists, etc.

In order to keep open these lines of communication between scientist and layman, a Society of Friends of the J.L.B. Smith Institute of Ichthyology, called ICHTHOS, was formed.

## *Why promote an interest in fishes?*

Fishes are the most diverse of all backboned animals and probably impinge on our lives in more ways than any other animal group. They are an important source of food and recreation and also provide agricultural feeds, oils and medicines. They are vital links in aquatic food chains and fascinating subjects for serious as well as casual study. People with an interest in fishes range from anglers, naturalists, conservationists, divers and aquarists to fish farmers, commercial fishermen and scientists.

## *Why was ICHTHOS formed?*

ICHTHOS was formed to provide an informal channel of communication between people interested in fishes. A variety of important research programmes are being conducted on fishes in the Ichthyology Institute and the Department of Ichthyology and Fisheries Science at Rhodes University. ICHTHOS would like you to know about their findings as well as those of other researchers, and would like to provide you with the opportunity to communicate your observations to them.

## *What are the benefits of joining ICHTHOS?*

1. Subscription to a regular newsletter giving interesting facts and figures on fish research, fishing, diving, aquarium keeping, fish farming, commercial fishing, etc.
2. Discounts on the purchase of books on fish published by, or sold in, the J.L.B. Smith Institute of Ichthyology.
3. Discounts on the purchase of books on natural history published by the Natural History Museum of London.
4. Notice of important books and articles on fishes.
5. Opportunities to purchase items of interest such as wall charts, books, T-shirts, sports caps and stickers, which are exclusively available to members.
6. Participation in the research of the Ichthyology Institute by providing fish specimens, field notes and donations.

## *How to join ICHTHOS*

For further information please contact:

ICHTHOS
J.L.B. Smith Institute of Ichthyology
Private Bag 1015
Grahamstown
6140
South Africa
Tel (0461) 27124
Fax (0461) 22403

# INDEX